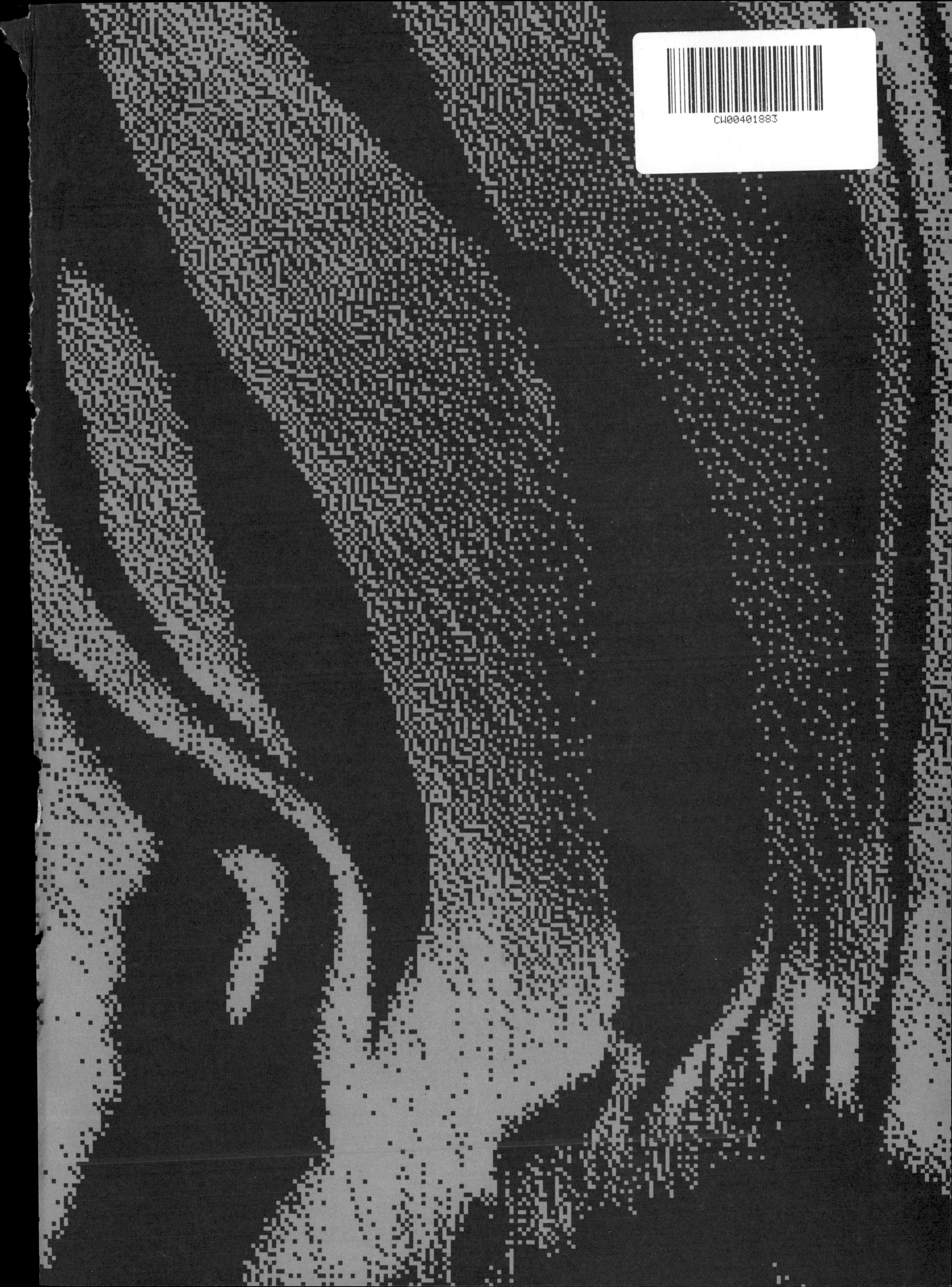

WILD TIGERS OF RANTHAMBHORE

YMCA Library Building, Jai Singh Road, New Delhi 110 001

Oxford University Press is a department of the University of Oxford. It furthers the University's objective of excellence in research, scholarship, and education by publishing worldwide in

Oxford New York
Athens Auckland Bangkok Bogota Buenos Aires Calcutta Cape Town Chennai Dar es Salaam
Delhi Florence Hong Kong Istanbul Karachi Kuala Lumpur Madrid Melbourne Mexico City
Mumbai Nairobi Paris Sao Paolo Singapore Taipei Tokyo Toronto Warsaw
with associated companies in Berlin Ibadan

Oxford is a registered trade mark of Oxford University Press in the UK and in certain other countries

Published in India
By Oxford University Press, New Delhi

First published in 2000

Designed by: Neelima Rao

ISBN 019 565 138 3

Printed at IPP Ltd, New Delhi 110 020
and published by Manzar Khan, Oxford University Press
YMCA Library Building, Jai Singh Road, New Delhi 110 001

WILD TIGERS OF RANTHAMBHORE

by Valmik Thapar

with photographs by
Fateh Singh Rathore

OXFORD
UNIVERSITY PRESS

CONTENTS

DEDICATION

This book is dedicated to those men and women who either sacrificed their lives while protecting Ranthambhore or gave some part of their lives to Ranthambhore and its tigers so that they could survive and enter the twenty-first century.

To the late Kailash Sankhala, M.Krishnan, S. Deb Roy, Avininder Singh, B.L.Meena, Hari Singh Meena (forester), Bhadulal (forest guard), Gopal Puri, Kalyan Singh, Ladulal (tracker).

To Shantanu Kumar, Salauddin Ahmad, P.K. Sen, G.V. Reddy, G.S. Rathore, Paola Manfredi, Malvika Singh, Tejbir Singh, Narendra Singh Bhati, Laila Tyabjee, Gajendra Jodha, Ujwala Jodha, Ashutosh Mahadevia, Ullas Karanth, Bittu Sahgal, Bhawar Singh, Rajesh Srivastava (R.F.O.), Daulat Singh (R.F.O.), Khem Singh Nathawat (forester), Brijbhusan (forester), Pushkar Maharaj (forester), Devi Singh (driver), Prahalad Singh (driver), Mahendra Singh (driver), Sayed Khan (driver), Balbir, Ghaffar, Hari Singh Kariri (forest guard), Sudarshan Sharma (forest guard), Ishaq Mohamad Barwara (forest guard), Gulam Mushtafa (head clerk), Nagina Mali, Shyam Singh, Hanuman Prasad, Jagan Gujjar, M.D. Parashar, Geoff Ward, Diane Ward, Peter Lawton, Sheila Lawton, Sunny Phillip and all the villagers of Kailashpuri and Gopalpur who decided to resettle outside Ranthambhore National Park in the 1970s, so that the Park could flourish.

AUTHOR'S NOTE

This book is a celebration of the tigers of Ranthambhore. It spans a quarter of a century, from around the time I first visited the Park and teamed up with Fateh Singh Rathore, when the tigers were totally nocturnal and elusive, to the times when they appeared throughout the day. In fact, so accustomed did they become to human presence that the jeep became like a bush or a patch of grass from which to ambush prey. The following pages, both in text and picture, are the essence of the finest moments that we shared with the tigers of Ranthambhore. Much of it has been published over the last two decades. I have edited the text in order to give the readers a glimpse into the secret life of tigers and an idea of the wonderful sights we had the good fortune to witness. In many ways the tigers of Ranthambhore rewrote the natural history of the tiger for the world, and some of our encounters were first-time observations of a predator that has always been clothed in endless layers of forest. The book is also a record of men like Fateh Singh Rathore and G.V. Reddy who took charge of an area and nurtured it with such zeal and commitment that the tigers responded in remarkable ways.

The population of the tigers of Ranthambhore has fluctuated much over twenty-five years: from 12–15 in 1974 to nearly 50 in 1989, to about 15 in 1992 and now at the end of 1999, climbing to 30. We have recorded over several generations, more than 125 tigers with our cameras. The process has been a great privilege for us and the most rewarding experience of our lives. We have also engaged in an enormous battle to ensure the safety of the tigers, from poaching and habitat destruction.

Selecting pictures for use in such a book has been a vast challenge. I had thousands of slides around me to choose from. I have tried to provide a visual insight into the tiger's reaction and response to the wildlife around it. The book is more about how the tiger looks at, and lives in, its environment—a very special and magical environment, full of the memories of man, a landscape dotted with the remnants of a historic past. It is less about how we as human beings look at tigers. I have selected only the rarest and most unique moments in the life of the tigers of Ranthambhore that we witnessed over twenty-five years.

Ranthambhore National Park is an island, fragmented, fragile and isolated, but it is a wonder of the world, a natural treasure that should live forever and it is upto all of us to ensure it does in the next millennium.

WILD TIGERS OF RANTHAMBHORE

The Seventies

If I track back to my first days in Ranthambhore, it is like another world, a moment in time twenty-three years ago when I boarded a train one morning for the small town of Sawai Madhopur in Rajasthan not suspecting that this first trip would change my life forever. Before that train journey, I had just directed two documentary films and my life was so city-based that few would have thought that the jungle would beckon. At the age of twenty-three I had a desperate feeling that something was missing in my life, there was an empty space. I felt that things were going wrong. I needed a break. Vanishing into the folds of Ranthambhore was a snap decision.

That train journey was one of the most difficult that I have taken. Somewhere instinctively I must have known that life was about to change. I had emptied out as if to await another experience. And it came.

> It is a strange six hours spent on the train, as it winds its way to Sawai Madhopur. While my eyes watch the flat dusty plains of Rajasthan, I think of my past and the delight of leaving behind a vacuous existence. But my apprehensions begin as I near my destination. There is no sign of a forest; only the same dusty plains devastated by the abuse of man over years. Sawai Madhopur certainly provides no clue to what might lie ahead. It is a typical district town, sprawled untidily around the railway track that seems to be its only reason for existence. My heart sinks a little, thinking of the endless train whistles that must surely frighten away any wildlife. As I step on to the platform, I feel the stares of people around me, all

seeming to question my presence. The roles seem reversed: I am the tourist attraction. I take the first horse-cart I can find and we jolt along to the Forest Department headquarters, through the dismal mess of concrete and brick structures that is the mark of every small town in India today. Smoke pours out of the cement factory nearby, a dirty grey cloud that takes over the sky. My hopes of ever seeing a tiger sink.

I met Fateh Singh Rathore, the Field Director of the Park, at the barrack-like office of 'Project Tiger'. He looked a strange sight with his luxuriant moustaches and Stetson hat. Had I arrived in the American Wild West? I asked permission to spend some time in the Park, and waited anxiously as he looked me up and down as if assessing whether or not the animals would approve. Fortunately I passed the test, and with great relief and a growing feeling of anticipation hired a jeep and driver and was on my way by the late afternoon.

Leaving the town, we followed a narrow metal road running parallel to a range of hills, and after several kilometres turned off onto a dirt track; but still there were few indications of what lay ahead, be it forest or wildlife. The road then turned sharply, and suddenly we were skimming along the rim of a deep ravine, bouncing and jolting over the stony track.

Suddenly the view changed. Below a sheer rocky cliff stood an ancient but massive stone gate that must once have been flanked by fortress walls, long since crumbled. It was a royal entrance to Ranthambhore, constructed to protect the domain of kings and surviving today to protect a treasure of equal if not even greater value. Water flowed from the gate through a marble cow's head, forming a pool at the entrance. Beyond the gate the air cooled, the vegetation thickened and the sounds changed. The chatter of birds mixed incongruously with the groan of the jeep. Cresting a rise, we saw the incredible sight of the Ranthambhore fort, grey and looming, extending upwards from a steep cliff face. The sky was a clear blue; the forest around, a dull green. The huge walls glinted in the evening sunlight, looking for all the world as if man had decided to chisel a bit of nature, the upper fringe of the rock, rather than disturb or fight it.

We were close to our destination. I tried to look around me, straining to see through the trees for signs of life, but my eyes were not yet accustomed to seeing in the forest. It is a skill not easy to come by. I could see old peepal trees and large banyans and I wondered how much they had changed since the days of Ranthambhore's glory. We were now winding our way below the massive fort, and as we crossed the last rise the terrain changed dramatically. The steep hills gave way to a broad valley, dotted with low hills and large expanses of water, the largest clothed almost entirely in giant pink and white lotuses. It was the pink of innumerable folk paintings, a pink that I had never before believed possible. It was too much to assimilate all at once, this mix of history, man and nature.

As we branched off the road I thought we were plunging straight into the largest banyan tree I had ever seen. Two of its hundreds of roots formed a natural gateway to Jogi Mahal, the forest rest-house, which decades ago had been the residence of a temple priest. Before the driver could even stop I was rushing up the steps, across a wide terrace and through the high arched doorway. There at my feet lay the lake of lotuses I had glimpsed earlier, with the waters of the lake lapping peacefully against the base of the rest-house.

> In the distance crocodiles lay sun-bathing, one with its jaw open, another swimming lazily in the water. Wild boar, chital and sambar are feeding on the lush grass growing on the banks of the lake. Some of the sambar are half immersed in the water, nibbling at the lotus leaves. Darters, herons, grebes and kingfishers fly around in abandon. Occasionally a large fish slaps the water. To the right of the lake are the remains of an old temple, and on one of the hills a guard post. Turning around, I look back the way we had come. The vast banyan tree and the backdrop of the fort fill the horizon with their imposing presence. It is a moment of hypnotic intensity and I suddenly feel exhausted. I know that Ranthambhore is going to mean something special to me.

The rest-house was simply appointed without the fuss or garishness of modern lodges. There was no electricity—Jogi Mahal was untouched by modernization. Soon, the forest staff collected around me; Laddu, the tiger tracker, Ramu, the wireless operator and Prahlad Singh, the driver. The chill of the night had descended and we lit a fire, the crackling of the firewood joining the nocturnal sounds of the forest. Crickets, nightjars, and owls had woken up to the dark. Small bats whizzed around and the large flying fox flapped above us. Laddu held forth on the tiger and his activity in the dark of the night, and Ramu talked about the ghosts left behind when wars ravaged the area. Prahlad added that even today a few spears could be seen embedded in the fortress walls. It was my first introduction to life in the ancient forest of Ranthambhore.

Without warning, this easy atmosphere was shattered by a loud booming noise lasting for nearly ten seconds. I took fright, and jumped, and amidst much laughter was told that this was the alarm call of the sambar. For them it was a normal night sound indicating the presence of a tiger or leopard. It would warn other deer that a predator was on the prowl and alert them to this danger. The call was repeated, echoing sharply off the fort walls. Torches were flashed around the thick roots of the banyan tree just in case there was something lurking around. Then a high-pitched shriek came from a little distance away. The chital, or spotted deer calling. The predator was moving. We were all on guard and I felt a shiver run down my spine as the tension mounted. After a while, the calls died away. The silence deepened and the forest relaxed again with the nocturnal sounds taking over. I felt part of the forest, even though only a small part. In cities one lives in isolation from nature and the forest suddenly puts you in your proper place. It had been a memorable evening. My eyes were heavy with sleep and I left the warmth of the fire to slink into my sleeping bag.

That first day in the forest, Ranthambhore entered my bloodstream. I will never forget the next twenty days spent learning about the ways and language of a forest and being initiated into a new world. It was on the last day of that trip that I saw my first

Ranthambhore tiger with the unmistakable glow of its striped coat; the powerful unhurried silent walk with which it confidently strolled down the middle of the road. The power and pure beauty of that moment cast a spell which was to become a driving passion in my life in the months and years ahead. For the next fifteen years Jogi Mahal was to become my home in the sun.

From 1976 to 1980 I visited Ranthambhore frequently but finding tigers or watching them was the most difficult thing to do. Looking back, it is like another era. Most people who visit Ranthambhore today do see tigers—you can see them on the first drive, in the park—and cannot comprehend that twenty years ago just seeing a pug mark or tiger scat was a great source of celebration. It was only in 1980 that I saw a tiger drinking water—tigers were invisible during those years, wary and frightened of man; when they appeared, they disappeared within seconds and sometimes you had to rub your eyes to believe that they were actually there! Most of those early years were spent driving at night in the hope of a glimpse of tiger or leopard. The days were for sleeping. The tiger was totally nocturnal.

Ranthambhore had several villages inside. There was human activity all day long inside the Park. The ban on tiger shooting was only six years old and tigers therefore retained their nocturnal clock that had protected it over centuries of exploitation by man. The sound of the gun and predation by man had taught it to move out only under the blanket of darkness.

And then Fateh Singh discovered Padmini with her cubs.

> It is early in January 1977. Fateh is sitting in a hide near the village hoping for some sign of the family, which had been active in the area during the morning. It is twilight, the first stars are appearing in the sky. The crickets and the nightjars start their chatter. An icy wind is blowing. Suddenly the stillness of the night is shattered by the shrieking death-cry of a buffalo. Immediately afterwards a brooding silence overtakes the forest. Not a sound in the dark.
>
> Fateh is moving off quickly towards the noise of the buffalo when the lights of the jeep catch Padmini head-on. Approximately 9 feet in length, with a beautiful coat and in the best of conditions, she stares at him and snarls. Suddenly, in a flurry of movement five young tigers run across the road, disappearing into the bushy country around.

It seemed as if on their mother's indication, the cubs had rushed into hiding. Padmini watched Fateh carefully and with measured tread walked away to her hiding cubs. Fateh moved out of the area, his heart beating faster at this first sighting of the complete brood. It was also the first sign of success after a patient wait of many days. The dawn brought Fateh back to the same area in the freezing cold. Early morning frost covered large stretches of the forest, his hands were numb driving the jeep, his eyes smarting with the chill, but there she was at the spot of the night before, sitting over her prey. To her right lay a full-size lame buffalo which she had killed the night before. Only a portion from the hind-legs had been eaten. None of her cubs were in sight. Padmini watched Fateh suspiciously. This was amazing behaviour for, during the day, tigers will normally find dense shelter to sleep in. But Padmini sat alert—no sleep for her, she was guarding the meat. She knew that every ounce mattered and she would have to keep continuous vigil to prevent it from being scavenged by vultures, jackals or hyenas. Because of the special situation her entire pattern of behaviour had undergone a change; she was spending the day alert instead of asleep.

In those years, the forest department would bait tigers by tying out young buffaloes. Tigers were so difficult to see that this was the only way across India to glimpse this striped illusion. We once used this strategy with Padmini's family to get a remarkable glimpse of tiger behaviour.

> The family watches us carefully as we push the unsuspecting young buffalo down the jeep and set it free. Then we move a little further up to wait and watch. This kind of thing requires total concentration. Your eyes are transfixed, you can't move, you can hardly breathe—it all seems to make so much noise. Padmini first leads her cubs towards the buffalo who senses imminent danger and tried to flee, but before it is out of range Padmini flashes in and disables its hind leg, making its escape impossible. After doing this, she goes and sits in the grass to watch.
>
> Now Akbar and Hamir move in to try and kill the young buffalo. But as they approach, it makes a limping mock charge and the young tigers retreat rapidly, unsure of their hunting prowess. Babar and Lakshmi are also inching forward now to join the fray. Suddenly, they encircle the buffalo and cut off any avenue of escape for it; but when any of the cubs come too close, the buffalo charges, using its horns very effectively. This whole 'game' lasts for thirty minutes, a long period of time during which Padmini never intervenes. The brave buffalo keeps charging all four tigers in different directions, forcing their retreat. During this time Akbar is the more aggressive and confident and it is he who suddenly leaps on to the buffalo's neck, bringing it down. Hamir then jumps on the hind portion and, between them, they finally kill it. The process takes time—the cubs are learning, but still have a long way to go.

A tiger is known for its perfect hunting technique; it kills quickly and instantaneously. Padmini could have finished the buffalo in a flash but preferred instead to lame it and then watch her cubs fend for themselves. These were lessons that her young tigers must learn. The most fascinating part however was to follow:

> Akbar and Hamir start eating immediately. They chomp on the meat for about twenty minutes until Padmini gets up and nudges them to leave. She then moves towards Babar and Lakshmi who are sitting a little way away, looking on. She nudges them as if to say, it's your turn now, but they do not budge, probably a little wary of our presence and the lights. Padmini goes forward and starts eating from the hind portion. We watch her for a short while, then leave soon after midnight.

Very few visitors came and hardly any saw tigers with or without bait. I remember once an American who sat out the night waiting for a tiger to take the bait—when the tiger did come it noticed the human intruder and roared in irritation. The American lost control of both bowel and bladder and the tiger ran off. The tourist had to be rushed back to a bathroom in the rest-house!

I saw many more leopards than tigers and on one remarkable day in February 1979, Fateh and I witnessed a unique encounter.

It is about 4.40 p.m. and the sun has lost its warmth. We stop the jeep on the brow of a hill overlooking the Semli valley. Soon we hear six or seven chital alarm calls, and speed towards their possible source. We find a few chital grazing but can see no sign of a predator. The chital stop calling, but suddenly Fateh spots a leopard's head in a patch of grass. The small, elegant head merges in perfect harmony with the grass, providing almost perfect camouflage as it pivots around, carefully watching the chital.

In the distance some sambar and a blue bull graze. Suddenly the leopard darts towards an unsuspecting sambar fawn. But in vain. To our immense delight, we discover that the hunter is a leopardess, and she now emerges again from cover accompanied by her two cubs. They are half her size, one slightly bigger than the other. It is the first time ever that I have seen a leopardess with cubs, and I become so engrossed in their antics that I do not realize the leopardess is no longer in sight. The two youngsters dash up a tree, leaping across from branch to branch. Then they scamper back down the trunk, and with a quick snarl at us, move off into a dry stream bed. As we turn the jeep around we see a large male hyena loping across the track. Another rare sight, as the hyena in India is solitary and hardly ever seen.

Delighted at our luck, we decide to return to base when we find the leopardess only a few metres away—sitting with an adult chital stag she has just killed. I cannot believe it. There we were, following the antics of the cubs in the tree in the tree while the mother, right behind us, has made her kill. I am sure the noise of the jeep must have helped her to stalk. Sad though we are to have missed the kill it still is a rare sight. She has just slit a layer of skin over the stomach—the characteristic method by which leopards start eating a kill.

The leopardess soon rises and moves off, while we pray that she will return with her cubs for the feast. The forest is still and silent. The hyena is approaching. It must have smelt the kill. Alert and watchful he comes up to the carcass and sniffs it, and then moves off some 30 metres or so. Suddenly, as if from nowhere, the leopardess is around him, hissing with rage as the animals circle each other. The stillness of the forest is shattered, and the valley of Semli echoes with the blood-curdling shrieks and moans of the hyena interspersed with the sharp coughs of the leopardess as they confront each other. But to our amazement the leopardess soon gives way and climbs into a tree. I am stunned. I cannot believe that a solitary hyena can get the better of a leopard. But so it was. The confrontation lasts late into the evening, but at no time does the leopardess launch a direct attack, or drive the hyena from her kill.

When I first went to Ranthambhore in 1976, Fateh Singh was in the process of shifting all the villages outside the boundaries of the Park. This was then a primary management objective of Project Tiger Reserves. The programme met with immense resistance from the villagers, many of whose families had lived in the forest for decades. It not only required hard work from the forest staff but also much understanding, tact and persuasion.

A variety of compensations were offered. The landless and the landed would all get an extra five bighas of land (about one-and-a-quarter hectares) in addition to what they originally owned. House sites and wells would be generously compensated for, and the resettlement would take place on agricultural land. The proposed new village would have a temple, a school and playground, facilities normally out of the villagers reach. Even so, the process was an uphill task as no incentive is enough when you are being uprooted from your traditional home.

Some of the villages did agree to move, and I remember those emotionally charged departures. A particularly vivid incident occurred in April 1977. It was early evening and all the district officials had assembled in the village of Lahpur to pay compensation to the villagers. The money had arrived and was ready for distribution. Suddenly I heard loud voices. The villagers had decided to refuse the money and not move out. Fateh took control of the situation. Huddled together around a fire, he and a group of angry and emotional villagers spent the next three hours in heated discussion while accusations were hurled back and forth. I listened spellbound. Every detail of the villagers lives, their perceptions and sensitivities, their children and wives and their future were discussed. I thought it was going to be hopeless, but Fateh's patience and tact won through. It was nearly midnight when the villagers agreed to accept the compensation and to move out, but only after their crops had been harvested. Their decision was tinged with a sadness and reluctance as this would be a radical change in the course of their lives. No longer would they live as a close community depending on each other for their daily needs. They would be a part of the mainstream, with all the benefits and problems of modern civilization.

Painful though it has been, I know that this shifting of villages has been a vital factor in restoring a healthy and harmonious ecological balance to the forest. The abandoned village sites provide a vivid example of nature's regenerating powers. Grasses and shrubs have overrun old fields, and mosses and lichens have carpeted the stone walls of abandoned dwellings. Some of the sites, such as Anantpura, Berda and Lakarda are now amongst the most likely spots in which to find tigers, leopards and bears. Deer and antelope have occupied the area as well and only a practised eye would be able to make out that there was a human habitation here only a decade ago.

Today, the forest throbs with activity, and with an abundance of plant and animal life. Large populations of sambar, who prefer to eat shoots growing in the water of the lake, rarely compete with the chital who prefer the grasses that grow on land. The langur monkeys have an abundance of wild fruit trees, and their stomachs are specially chambered to allow them to digest the leaves. The blue bull is seen grazing away from the lake areas and the delicate cinkaras, who obtain water from the plants they eat, frequent the remote hard-ground plateau and the hills of the interior. Sloth bears thrive on wild fruits and wild boars dig up the ground in search of roots.

I believe that the only reason the tiger is alive in Ranthambhore at the eve of the new millennium is the sacrifice made in the late 1970s by the forest dwellers who moved out of the forest so that the tiger could flourish. Tigers and human beings cannot share the same tract of forest. One of them had to go. Fortunately in Ranthambhore, the humans agreed to relocate.

Their relocation led the tiger into rewriting its own natural history. The 1980s in Ranthambhore were some of the most remarkable times for the tiger when new facets of its behaviour were continuously being recorded.

The Eighties

In the eighties in Ranthambhore, the tiger suddenly became more visible. It was a gradual and surprising change. But exactly what was happening? The tiger population had certainly increased, but I had to analyse why the tiger was more relaxed and confident, fearless of the human presence. There seemed a steady and continuous increase in tiger activity throughout Ranthambhore's dry, deciduous habitat and especially in the more open areas and on the edges of the lakes. Tigers had begun appearing everywhere, and their increased activity was not restricted to the night any longer.

On one occasion I spotted a tigress stalking the edge of a lake in mid-afternoon, when the temperature was 45°C. She was carefully watching a group of sambar grazing, almost immersed in the still waters of the lake. Suddenly she took off at great speed into the water, first wading and then swimming in relentless pursuit of the startled deer who rapidly fled to the far corners of the lake. The tigress then proceeded to sit in the middle of the lake, completely relaxed—astonishing behaviour for a formerly nocturnal hunter.

I also remember sitting on the terrace of Jogi Mahal sipping a cup of coffee, reflectively watching an idyllic pastoral scene—a herd of spotted deer grazing at the edge of the lake—when suddenly a tiger bounded out, killed a doe and rushed back into the tall grass with her. An even more exciting observation made from this balcony was a tigress attacking a group of five spotted deer at the edge of the lake. She leapt on one and killed it. The other four deer fled into the water in an attempt to cross from one side of the lake to the other. Before they could even reach halfway each of them were plucked from below and in sudden splashes vanished into the mouths of crocodile. One tiger kill had led to four other crocodile kills. Jogi Mahal's balcony is one of the finest observation posts I have ever encountered in India.

In these memorable years, a new generation of tigers were being born to mothers who had never known man's aggression. The mother no longer taught them to avoid man and the tigers were coming into their own. The tiger's whole perception had changed: it was becoming fearless.

Fateh Singh Rathore and I had some unique encounters and I will try and reproduce below, from our old records, a little of the phenomenal excitement of those times.

Very rarely does one come across records or evidence of fatal interaction between tigers in Ranthambhore. Maybe these do occur, but one can seldom prove it as the forest destroys all evidence in its natural cycle, burying it in the earth. Ranthambhore in all these years provided just a few examples of mortal combat between tigers; we tried to piece together the events from evidence we discovered the morning after it happened.

There was a full moon on the night of 10 November 1981 when a tigress and two cubs appear to have been walking down the Lahpur valley nearly 20 kilometres from Jogi Mahal. She must have spotted an adult male tiger walking in the opposite direction. Indications existed of the cubs scampering away. The tigress seemed to have continued towards the tiger and there were marks where both sat down in the middle of the road. They must have then risen and gone to sit in the sandy part of a nearby stream bed. Obviously at this moment the tigress was doing her best to be affectionate with the male and bid him a rapid farewell before any interaction was possible between him and the cubs.

But it did not work. The cubs seem to have attempted scampering back to their mother, probably finding the insecurity of separation too much to take. At this moment there must have been havoc, and some incredible vocalization was even heard in a guard post some 2 kilometres away. It appears that the male moved in a flash towards the cubs, and the mother was forced to take quick action. With a leap and a bound she attacked the male from the rear, clawing his right foreleg before sinking her canines in and killing him. It was an amazing example of instinctive reaction: a tigress killing a prime male tiger to save her cubs from possible death. The male must have been caught completely by surprise and just succumbed. Later the tigress proceeded to open his rump, and eat off his left hind leg. Tiger eating tiger: this was a rare example of a fatal interaction between them.

A chilly November morning 1982. We were about to witness the rare sight of several tigers interacting over a natural kill:

> We are driving around the third lake when in the distance we see a frantic tracker on a bicycle, gesticulating wildly as he approaches us. He shouts, 'There is a tree full of crows and I have just seen a tiger eating on a nilgai by Rajbagh.' We rush off and sure enough come to a tree with nearly fifty crows on it. Below it sits Padmini with three 14 month-old cubs around her. Nearby lies the carcass of a huge adult male bull, which must weigh at least 250 kilograms, possibly more. The carcass is far too heavy for her to move and Padmini is nibbling at the rump, a small portion of which has been eaten. The two cubs sitting behind her get up in an attempt to approach the kill, but as they come close she rises, coughs sharply and slaps one of them across the face. The cub submits, rolling over on its back, and settles down near her restlessly while the other cub starts to eat from the rump. Padmini seems to be saying, quite clearly, 'One at a time.'
>
> At 7.30 a.m. Padmini gets up, grabs the nilgai carcass by the neck and tries to drag it away, but its foot gets stuck in a forked tree root. She settles down to eat some more, and half an hour later tries again– this time dragging the carcass about 8 metres. Now she permits the second cub to eat. The third cub waits some 30 metres away. Crows chatter incessantly and a group of vultures circle while others sit on a nearby tree. The crows fly in and around in an attempt to pick up scraps of meat but twice Padmini charges them. A single vulture flops down but in a flash he too is charged and takes flight.
>
> Soon after this, Padmini drags the carcass about 10 metres farther up the rise of a hill. We follow quickly as the terrain is easy here and accessible to a jeep. The great advantage of Dhok forest is its excellent visibility, and scanning the area we now see Laxmi, Padmini's female cub from her litter of 1976. We now have five tigers spread out at different distances around the carcass, Padmini and Laxmi closest to it. There is much getting up and sitting down and Laxmi twice marks the trunk of a tree. Now Padmini gets up, sniffs a tree, spray-marks it and walks towards

her nearest cub, nuzzling it briefly. She then turns around and walks past Laxmi, snarling at her before grabbing the neck of the nilgai and pulling it farther up the hill. Two of her cubs are sleeping in the distance, but soon all three rise and circle out of sight, Padmini and Laxmi are lying side by side near the carcass.

A few minutes later Padmini gets up and walks down the slope of the hill towards the lake. Laxmi moves towards the carcass and starts to feed. It is 11.00 a.m. We follow Padmini as she moves towards the edges of the second lake and flops into a patch of water. She drinks a little, rolls around and gives herself a good soaking. After twenty minutes she rises and moves into the long grass. Sounds of snarling and growling are heard and the faint outline of two tigers can be seen moving in it. Padmini soon emerges from the grass and sits by the roadside.

We leave her there and go back to the nilgai. Laxmi is eating but surprisingly there is now another adult tigress sitting nearby. It is Nick Ear, a female from Padmini's second litter. At noon Padmini appears from the rear, marks a tree and moves towards the kill. She and Laxmi cough at each other. Padmini sits, and snarls at Laxmi who moves off towards Nick Ear and settles down on her side to sleep. Padmini also dozes off but with a watchful eye on the crows perched on the branches. At 12.30 p.m. the dominant cub returns from the lake-side and sits at the kill, nibbling at the fast diminishing rump while Padmini watches alertly.

At 1.00 p.m. the cub moves off and rolls on his back before flopping on his side to sleep. Padmini gets up and chews on the carcass for some fifteen minutes, then she too moves off again towards the lake. Laxmi rises and settles near the carcass. At 2.00 p.m. Padmini returns to her original position and a grimacing Laxmi moves back. From our rear two more of the cubs emerge. There are now six tigers in front of us.

But more is to come. About 45 metres away, amidst growling and snarling, yet another female, Nasty, appears growling continuously as she approaches the carcass. She is the seventh tiger on the scene, but this time one without any kin-link to Padmini. Time seems to stand still. Then at 3.00 p.m. we spot a large male walking along the edge of the hill. He sits down some distance away, and after carefully studying him with our binoculars we discover that it is Akbar, the dominant male of Padmini's first litter. And behind him, with thudding hearts, we see yet another tiger, but at this range we cannot identify it.

Fateh and I look at each other in near hysteria. Nine tigers surround us at varying distances from the nilgai! Hardly daring to breathe, our eyes switch from tiger to tiger. None of them are eating. Four are sleeping, two are grooming, and two are watching the kill. One is on its back, paws in the air. Padmini seems to be completely in command of the group and is obviously the one who killed the nilgai. Soon after 3.30 Nick Ear moves towards the kill and eats for some twenty-five minutes. She then moves away and Padmini's dominant cub arrives to feast. Laxmi also decides to eat but a sharp snarl from Padmini sends her off. After the cub has eaten for fifteen minutes he moves off, and Laxmi comes for her turn.

So far we have seen five different tigers eating, but only one at a time and all strictly controlled by Padmini. There seem to be one male and two females in her third litter. Laxmi now moves off towards a dry stream bed from where chital alarm calls are sounding. Perhaps her own cubs are hidden there: so far we have seen no sign of them. She calls six times before moving out of sight. The time is 4.30 p.m. and the shades of evening colour the forest. At 5.20 p.m. we leave the hill after some ten uninterrupted hours of quite extraordinary observation.

The next morning, on returning at 7.00 a.m., I find five tigers, Laxmi, Padmini and her three cubs, sitting around the kill with just the head and ribcage of the nilgai left. After chewing for forty minutes they disappear over a rise and into a nallah. Now only the crows remain, picking at bits of bone.

I have never come across a description of a scene like the one we have just witnessed—where nine tigers have fed on the same carcass, entirely controlled by a dominant female. It would appear that Padmini had made the kill but as several other tigers were also present in the area she took a decision to share the carcass. As it happened, all except two of the tigers were sons and daughters from her various litters. But not once did she permit two tigers to eat together, thereby preventing the conflict that could have arisen.

To find nine tigers around a kill is a rarity. I think the fact that seven were related signifies the possibility of strong kinship links among tigers and that these may be sustained over long periods of time. I think tigers easily recognize another, be it brother, sister or mother. A great deal more observation is needed for conclusive evidence on kinship links and their role. But this example shows that tigers can congregate without conflict around a natural kill—and with a female in full control of the feeding process.

> November 1983. We receive information that five tigers had slipped into the area of the lakes along one of the stream beds. We begin circling the meadow. A few chital are grazing in the centre, but of the tigers there is no sign. Our excitement is high. This could be the adult group we know had been roaming the area for some months. Switching off the engine we suddenly hear three frantic chital calls and then a choked squeak. Fateh, convinced that it is the death cry of a chital, moves to the spot. Just ahead of him, a tiger sits on its haunches, panting heavily.
>
> We move on, and as we cross a patch of grass we encounter two more tigers. One is sitting with its paws hugging the carcass of a chital doe while the other watches alertly, moving a few steps forward. The first tiger emits a low growl and then with a loud 'woof' charges the second tiger. Both rise briefly on their hind feet 'mock boxing' each other, but soon the second one rolls over on its back as if in submission. The first returns to the chital kill, followed closely by the second. Fateh notices that both are adult females. Woofing, coughing and growling, the two tigresses start pulling at the carcass from both ends in a regular tug-of-war, but the first tigress, pulling at the neck, seems to be gaining ground. They move some three or four metres in this strange manner until the first, and obviously more dominant of the two, lets go of the carcass and leaps at the second tigress. The latter immediately releases her grip and the dominant female again sits over the carcass as if hugging it. Low but ferocious growling emerges from her throat. Paws extending over the chital kill, she is vigorously asserting her rights over her prey.
>
> The second tigress is now crawling towards the kill, burying her head alongside the paws of the first until her head is close to the carcass and near the neck of the first tigress. The first tigress now snarls viciously at the other, but it has no effect. Amazingly, they continue this for thirty minutes without feeding. The growling, coughing and snarling rise to a crescendo. I had never heard such a variety of tiger sounds before. It is aggression only through sound—at no time did the two animals attempt to actually injure each other.

In March 1984 a stroke of luck led us to a recent tiger kill near the Kachida valley, and over a period of three days we were able to watch a complex interaction, over the kill, between Laxmi and the male tiger Broken Tooth.

> Day 1, 1.30 p.m.: I am driving down a track leading into the Kachida valley. Something is bothering me, and for no reason other than an instinctive feeling, I take the jeep off the track and cross country. In the distance a group of tree-pies attract my attention and I set off to investigate. Interpreting the sound of tree-pies is a vital clue to wild tigers. And sure enough, in a bend of a rocky stream bed, I find Laxmi with recently killed sambar hind.
>
> 1.45 p.m.: Laxmi rolls onto her back, totally relaxed, having eaten a small

part of the sambar's rump—the choicest part and nearly always the first part of the carcass eaten by a tiger. Twice while dozing she is disturbed by thieving tree-pies which she drives off with snarls and a couple of mock charges.

3.20 p.m.: Laxmi rises, sniffs the carcass then paces round it. She attempts to move it by the neck, but then instead tries to cover it by scraping leaves and stones over it with her feet. Tigers tend to do this out of instinct. I have never seen it done well enough to cover a carcass but the action probably does help to hide the meat from crows, tree-pies and vultures. In this case she even attempted to push large-sized boulders on it.

5.40 p.m.: Laxmi feeds again, this time opening up part of the abdomen and devouring the sambar's intestine. After eating she settles down again to rest. A solitary Egyptian vulture flies slowly overhead, but like the tree-pies he will get nothing; Laxmi dozes, lying almost on top of her prey.

Day 2, 6.35 a.m.: On returning the next morning we are amazed to find a large male tiger sitting over the kill. Laxmi is 20 meters away, looking towards him. The new arrival is Broken Tooth, a resident of the Kachida valley. One of his lower canines is broken but otherwise he is a prime male. The sambar carcass has been greatly reduced. It is obvious that both tigers have feasted. Very rarely do you find a second tiger on a kill and such encounters add to understanding the secrets of the tiger's life.

7.00 a.m.: Laxmi looks towards the carcass, but each time she does so a soft warning snarl from Broken Tooth discourages her. He has appropriated her kill. Perhaps there had been some aggressive vocalization from Laxmi initially, but now she lies submissively at a distance.

8.45 a.m.: After a short nap, Laxmi rises and moves away. Broken Tooth lies with his head resting on the carcass: tigers seem to like being physically 'in touch' with their food. Between now and 11.00 a.m. Broken Tooth dozes—but five or six times he is disturbed by tree-pies trying to scavenge scraps of the kill.

3.00 p.m.: It is now very hot and Broken Tooth rises, tugs at the sambar carcass, then lets go again. He turns and walks off towards a small water-hole. The temperature is nearly 40°C and in that heat quenching thirst and soaking the body are essential activities for a tiger.

3.10 p.m.: Broken Tooth returns, dripping and refreshed. He drags the carcass a short distance and settles down to eat. Soon we hear the incredible 'sawing' sound of his canines slicing through the sambhar's hide, followed by the crunching of bone.

4.00 p.m.: There is still no sign of Laxmi returning. Broken Tooth has spent the afternoon dozing and occasionally chewing on one of the sambhar's hind legs. He settles down to sleep and we head back to the rest-house.

Day 3, 6.45 a.m.: Returning to the stream bed we find Broken Tooth still 'in residence'. The sambar is almost eaten. Only the head, forelegs and part of the ribcage remain. The tiger's belly is full and he looks much bigger. He sprawls beside the remains, a picture of contentment.

10 a.m.: Broken Tooth is roused from his doze by the appearance of another female. She has a wound in her rear flank and is obviously hungry. Broken Tooth snarls viciously, and the tigress retreats out of sight. It is probably a young tigress that has been swiped on her rear while competing with another tiger over food.

4.00 p.m.: Broken Tooth strolls down to the water-hole and soaks himself, staying there for nearly thirty minutes. We decide to leave. About 2 kilometres along the track we come across Laxmi, sitting beside another nallah. Is she planning the strategy for her next hunt, or thinking of the fine deer that was snatched from her by the big male tiger? All her hard work, so that the big male could feed.

And then there was the legendary tiger Genghis, a large male who exploded across Ranthambhore lakes between 1983 and 1984. Never again has a male tiger spent so much time as Genghis did on the lakes—and he was unique. I had over many

months of watching him been able to predict his movements and the routes he would use and seven times out of ten could find him around the lakes. To me he was the great thinking tiger of Ranthambhore. My records of March 1984 contain one of a typical 'Genghis day':

> Genghis moves into this patch of long grass at about 11.30 a.m. after consuming the remains of a wild boar piglet he had killed the previous evening. He sleeps in the shade of the grass until 2.30 p.m. when a group of sambar appears on the shore and starts to move towards him.
>
> As the sambar draws close, Genghis erupts from the grass in a headlong charge, scattering the panicky deer. The sambar flashes past our parked jeep and only at the last second does the charging tiger even see the vehicle. Slamming his forepaws into the earth he brakes to a halt, then veers away snarling in frustration. At close range, head-on, it is a display of breathtaking power, strength and speed.
>
> Malik Talao, 5.20 p.m. Genghis stands motionless in the tall grass at the lake edge, deciding which of the sambar would be his main target.
>
> His eyes settle on his target. Genghis has started his charge, rushing diagonally through the grass towards the sambar. The deer have seen him and with tails raised, calling in alarm, they flee farther out into the lake. It is just what Genghis wants. His diagonal run has cut off any chance of the sambar escaping onto the shore, and now they are forced into deeper water, confused and in total panic.
>
> With a mighty leap Genghis launches himself towards the water of the lake. The sambar are frantically trying to flee but the weight of the water hampers their movements. There is complete chaos amongst them.
>
> Crashing through the water amidst sheets of spray, Genghis swerves towards his chosen target in an attempt to cut it off. His power and speed in the water is astonishing. He has selected a young fawn as his target. Its mother realizes and turns in anguish, knowing that there is little hope of survival for the fawn, yet reluctant to desert her offspring.
>
> Genghis is now closing fast, pounding though the water with powerful strides. The sambar hind rushes away: she has given up. The tiger's paw smashes down on the helpless fawn with such force that fawn and tiger disappear beneath the water. Only Genghis' tail is visible. This is one of the most remarkable recorded sights of the world's most powerful predator in action. Beneath the water the fawn is in its death throes as Genghis' canines close in a vicelike grip on its neck. The sambar hind watches in distress.
>
> Genghis wades ashore, carrying the fawn in his month flicking water from his tail as he heads for the cover of the grass thicket to feed in privacy. The chase, from his emergence from the thicket to his disappearance back into it, has lasted barely two minutes.

Genghis was a unique tiger and ruled Ranthambhore's lakes for one season—he attacked the crocodiles, snatched carcasses from them, frequently swam across the lakes and could heave out sambar carcass from deep in the water. He vanished within nine months and I never saw him again. But his records of predation in the water remain unique in the body of tiger literature across the world. He also revealed how individualistic tigers are in their temperament. He would not tolerate human observers while he was eating and would come out at full pelt charging the jeep. His aggression was intense.

In March 1985, I had my first real insight into family life when I discovered Laxmi with a brood of three tiny cubs. For nine years we had waited for such a moment. It was like a dream come true.

I stop the jeep. It is time to wait for sounds of alarm. In the distance two sambar hinds move gingerly away, tails half raised. A chital looks sharply towards the forest. It is motionless. My eyes are unable to pick out anything. The shrill alarm of a peacock breaks the silence. Another peacock picks up the call. After a few seconds the alarm call of the chital pierces my ears. Frenzied and frequent calling now surrounds me. Quietly I watch the forest. It seems as if the tigers are walking towards the vehicle track. Suddenly shades of tan and black emerge from the dull yellow of the forest. Laxmi appears with three tiny cubs. One of the cubs jumps across the road. It looks about two and a half to three months old. I can hardly believe my eyes. It is my first glimpse ever of cubs this size. Laxmi settles down on the track for a few minutes. Her cubs look at me furtively from the cover of a bush. She soon rises and paces leisurely into the forest, followed by three scampering cubs. They move towards a network of ravines and disappear from sight. For me it is a dream come true. I rush back to base, heart pounding with excitement. In near hysteria I tell Fateh what I have seen. We sit down to plan strategies for the following weeks.

The next morning I find Laxmi sitting in a grass patch 10 to 15 metres from the forest track. Her three cubs surround her. One nuzzles her face, another rests against her back, the third watches us curiously. Very tentatively it moves a little towards us before rushing back to the security of Laxmi. The cubs now turn their attention on each other, leaping into the air and knocking into each other. They then dash towards Laxmi. She licks one of them thoroughly, then lies on her side to suckle them. All three soon find the right teat and feed, stimulating the flow of milk with their tiny paws. For fifteen minutes I watch this remarkable spectacle. I have never seen such a display of love and warmth, such evidence of the strong bond between a tigress and her cubs.

For us to witness mating tigers was like another dream. Fateh was the first to get lucky. From 6 March to 9 April 1985 Kublai and Noon spent much of their time together. It was the preliminary phase of their courtship, and much of their interaction was over food. We watched the pair endlessly around the lakes and if I look back, that period of time was sheer delight.

10 March, 1.20 p.m.: Noon drags ashore the carcass of a sambar hind she has killed in the shallows of Padam Talao. She has retained her vicelike grip on the throat, and the sambar's body trails between her forelegs and beneath her body. Once ashore she pulled the carcass into long grass 100 metres back from the lake in order to feed. I was always amazed to see how Noon had become predator in the water after Genghis' departure. The success rate of killing in the water was definitely higher.

4.00 p.m.: Noon returns to the lake to quench her thirst and to soak in the cool water. Tigers often do this particularly after making a kill, and during feeding, particularly in hot weather. Noon is even fonder of the water than most and is often seen partly immersed.

4.20 p.m.: Noon suddenly rushes from the water and charges back to her kill, clearing a deep muddy wallow with one 5 metre leap. She had probably seen a greedy tree-pie heading for the meat!

7.30 p.m.: A deep resonant 'aaoon' reverberates across the lake, soon followed by another. Kublai is moving towards Noon. Tiger sounds echo off the wall of the fort for another twenty minutes. Has Noon invited Kublai to the feast? Or has the

male's appearance some days ago stimulated the tigress' oestrus cycle?

11 March, 6.45 a.m.: Noon is sitting some 10 metres outside the grass and we suspect that Kublai has taken over her kill. She rises and moves tentatively towards the grass, but retreats again submissively in the face of a series of low rumbling growls. Later in the day she tries to stalk another sambar but is unsuccessful, and for the rest of the day is kept away from the kill by Kublai who remains in the grass, not even coming down to the lake to drink.

During this preliminary courtship, Kublai annexed at least four of Noon's kills, and only once was she able to snatch part of one back—while Kublai was taking a drink at the lake. He was firmly established as the dominant male and Noon's food intake suffered considerably.

9 April 1985 saw Fateh's seventeen years of painstaking observation rewarded when he was able to watch, and record on film, the culmination of Kublai's month-long courtship of Noon. In 1976, when I first visited Ranthambhore, no one would have believed that tigers would be seen mating around the lakes within the decade.

5.30 p.m.: Kublai and Noon have already mated twice, but partly hidden in the dense grass. They now laze at the edge of the thicket indulging in protracted love-play. Noon nuzzles Kublai, they rub heads, and Kublai's paws touch the tigress's chest. She is standing while he remains sprawled on his side as she tries persistently to arouse him. Fateh has to actually rub his eyes to believe the scene in front of him.

5.37 p.m.: Kublai leaves the grass and strolls off round the lake followed by Noon who at one point overtakes him and leads the way. But their attention is caught by a sambar carcass floating in the water. They make no attempt to rush in and drag it out, but the presence of food so near brings their walk to an end and they remain on the shore nearby.

5.45 p.m.: Noon moves quickly to Kublai's side and with a brief nuzzling movement along his head and neck she brushes her flank against him seductively. By this time Noon was forcing a response from the male every fifteen minutes or so in her eagerness to mate. Fateh found the energy levels of the tigers and their single-minded obsession with each other remarkable.

6.12 p.m.: Yet another period of copulation takes place, despite the fact that Kublai has spotted Fateh and is watchful. This time he does not grip her neck, and Noon is not aggressive; she lies at full stretch, her chin almost on the ground. Tigers copulate with some frequency and in this case the pair mated eight times between 5.20 and 6.48 p.m. A new chapter in the life of the tiger has opened for all of us.

And even more startling was in April–May 1986 when for the first time we had real glimpses of the father in the family unit and our records of this were the first in the world of tiger behaviour.

At four in the afternoon Kublai lazily ambles towards the pool and slides into the water, hind legs first, soaking himself completely, leaving just his head visible. Tigers don't like water splashing in their eyes and most of them enter water backwards.

About twenty minutes later Nalghati follows and they both laze around in the water. Minutes later both my heart and Fateh's must have missed a beat. The male cub walks quite nonchalantly towards the pool, not a flicker of surprise or fear on his face, circles the two adults and enters the water near where Kublai is

stretched out. Soon, following her brother, the female cub walks to the pool, entering the water to sit on her mother's paw. Nalghati licks her face. Fateh and I cannot believe our eyes—the tranquillity of the scene is extraordinary. One big happy family: Nalghati, Kublai and two five-month-old cubs, all in close proximity, soaking themselves in this rather small pool of water. They lap the water at regular intervals. In half an hour the male cub rises, quickly nuzzles Kublai and leaves the pool. The female cub follows him and they play, leaping at each other, slowly drifting towards a tree, clambering up the branches to play a game of hide and seek amidst the foliage. The two adult tigers watch. Soon Nalghati leaves the water and disappears into the forest. The cubs continue to play with each other under the protective eye of Kublai. At dusk, Kublai heaves himself out of the water and moves towards the cubs. The cubs rush up to him. He licks one of them.

When we leave, Kubali is sitting a metre or so from the two cubs. We have witnessed what must be one of the most closely kept secrets of a tiger's life. It is the first photographic record of a resident male associating with a tigress and her cubs in his range.

The next morning Fateh decided to go towards Semli, acting purely on instinct. I went to Nalghati but was unable to find the tigers. They must have been around since the crows and vultures were still perched on the trees. When tigers are away from their kill, vultures and crows leave their perches in the trees, dropping to the ground to consume the remnants. If they are near the carcass, the tigers have been known to kill the vultures as they land, with a smack of their powerful paws. So the birds waited patiently on the trees. The dense forest cover hid the tigers. I returned to base early and Fateh rolled up an hour later, beaming. I knew immediately that something exciting had happened.

It is 1 May 1986. Just at the edge of Semli, in the gorge of Bakaula, Fateh finds the Bakaula male and Laxmi. They are sitting on the vehicle track facing each other. On both sides of the track are thick groves of jamun, cool, lush and green. There are pools of water of various sizes nearby. Laxmi rises briefly and nuzzles the Bakaula male before moving a little way ahead to lie down on her side. A stork-billed kingfisher calls near the water. A pair of Boneli's eagles circle above a cliff.

This tranquil scene is disturbed by the distant sound of a rolling pebble. Both tigers become alert. Laxmi moves stealthily towards the sound. The Bakaula male sits up expectantly. A sambar deer shrieks in alarm. Laxmi has disturbed it. Tail raised vertically, the sambar carefully walks down an incline. Laxmi is too far away to attack. The sambar's path is taking it unknowingly towards the Bakaula male, who crouches, muscles tense. The sambar approaches the vehicle track. The male tiger takes off like a bullet. Six bounds and it leaps on to the back of the sambar, bringing it crashing down. Quickly it transfers its grip to the throat. At the same instant a group of noisy tourists arrive, stunned at seeing a tiger choking a sambar to death bang in the middle of the vehicle track. But the male is disturbed and walks off behind a bush. The sambar is not quite dead. It twitches with small spasms of life. Laxmi arrives.

Comfortable in the presence of jeeps, she grips the throat of the sambar for a couple of minutes, ensures that there is no life left in the deer and starts the tedious process of dragging the 180 kilogram carcass away, a few metres at a time, into thick cover. The Bakaula male watches her carefully. The carcass is now some 15 metres inside the jamun grove at the edge of a small clearing. The tiger moves towards it. So does Fateh. An amusing scene confronts him. The male tiger, with his forepaws on the sambar's rump, has a firm grip on one of the hind legs. Laxmi has a firm grip on the throat. The carcass is stretched between the two tigers. A mock tug-of-war ensues as each tries to pull the carcass a little towards itself. Both tigers emit low-pitched growls, interspersed with herculean tugs at the carcass. Then, with a sudden burst of energy and strength, Laxmi yanks the carcass some 4 metres away with the Bakaula male astride its rump: a remarkable feat, as sambar and tiger together must weigh about 450 kilograms. But it exhausts her and she lets go of the throat. The male quickly pulls the carcass out of sight.

Laxmi strides off. Fateh follows. She enters a dry stream bed that leads to her den. She starts to call and is greeted by birdlike squeaks from her cubs. The complex and elaborate language of the tiger resonates through the atmosphere. In minutes Laxmi returns with the cubs running around her in circles. One of them runs between her legs and tries to leap over her back. The other two are frisky and jump up the trunks of trees before slowly moving to where the carcass lies. The cubs have already learnt the art of sniffing and they follow the drag marks of the carcass. They seem quite relaxed, as if this wasn't the first occasion that they were going to share a feast with the Bakaula male. The male cub suddenly sniffs the spray mark of a tiger on a bush and wrinkles up his nose in the gesture of 'flehmen'. Soon they all disappear out of sight to where the Bakaula male and the carcass lie. Within the last two days we have twice seen a remarkable facet of the family life of tigers; the resident male in the role of father.

One warm day in March 1987 I arrived at the Semli water-hole to find the three cubs resting in the cool of the undergrowth. The female cub moved towards us as we arrived, in her normal way, walking past very close to the jeep. There was no sign of Laxmi. The cubs lazed around for nearly an hour and at four o'clock one of them suddenly became alert. It darted off to far side, followed by its siblings. The forest exploded with the sound of purring as the cubs exhaled in great bursts, probably indicating joy. We followed to find the cubs rubbing their flanks against Laxmi. All four tigers purred incessantly, as if orchestrated, as the cubs licked, nuzzled and cuddled their mother. I have never ever heard anything like it.

The purring continues for nearly ten minutes, as all four tigers walk towards us. The cubs rub their bodies against Laxmi, expressing their delight at seeing her. They move to a water-hole and quench their thirst, then Laxmi moves some 20 metres away to rest in the shade of a tree. The cubs return to their original positions.

It is nearly five in the evening and we decide to watch from a distance, hoping that some deer will soon come to the water-hole to quench their thirst. At six o'clock a group of fourteen chital emerges from the cover of the forest and cautiously approach the water. Laxmi is suddenly alert, watching intently. The deer have not seen her or the cubs. The cubs lie frozen, knowing the slightest movement will give the game away. Most of the deer have their tails up. They are suspicious, but they need water. Slowly, step by step, they approach. They are now between the mother and her cubs. A perfect situation for Laxmi. She crouches, moves forward some 3 metres on her belly as if gliding along the ground. Then one chital alarm call pierces through the evening.

Laxmi moves in a flash, her cubs sprint from the far side and in the panic and confusion of the moment a fawn gets separated and flees towards Laxmi. Laxmi pinions it between her paws and grabs the back of its neck. The fawn squeaks and dies. A stork-billed kingfisher flies away from its perch, its blue wings glinting in the late evening light. Picking up the fawn, Laxmi carries it some metres away.

Her cubs move in, hoping to feast on a few morsels. Laxmi drops the fawn to the ground and settles on it, covering the tiny carcass with her paws. She turns and snarls viciously at the approaching cubs. One of the female cubs moves off but the male and the other female settle down to face their mother a metre away. Both cubs start a low-pitched moaning sound, which I have never heard before. It soon turns into a wail as if they were begging for the carcass. Laxmi snarls and coughs sharply at them. The male cub rises and moves towards her but she growls, picks up the carcass and settles down again some 3 metres ahead, the fawn between her paws.

All three cubs settle down around her, moaning continuously. She snarls back and this continues for fifteen minutes. Suddenly two of her cubs 'cannonball' into her and all three tigers go rolling over in a flurry of activity, but the male cub snatches the carcass expertly and rushes away with it, followed by one of his sisters. Laxmi sits unconcerned and proceeds to groom herself. The male will not tolerate his siblings on the carcass and they return to Laxmi. They watch the male cub eating. They wait patiently for forty-five minutes until the fawn has been consumed. Then, they all move off. I have found that when the prey is tiny, the dominant cub asserts his right to eat most of it. He shares only when the kill is large.

Tigers are basically silent animals and vocalization is rare. In twenty-three years of tiger watching in Ranthambhore I have never heard purring the way I did on this occasion.

Late that month I found the large Bakaula male vanishing into the thick and cool jamun groves in Bakaula. I felt a great urge to call out to him and in my best tiger voice I mimicked the tiger's roar. Within a minute this huge and enormous male tiger came pacing out of the bush and sat 15 feet away looking at me as if I was mad. I looked at his eyes and realized that he was irritated by the disturbance and quietly I moved away. Obviously, there are times when you can attract tigers by 'calling out'!

In March 1987 I also had a memorable morning with Noon and her cubs, watching them over two natural kills within an hour of each other. We had stopped near Rajbagh and the barking of the langur monkey from the side of the lake spurred us on. The sun was just rising over the hill. Amidst a cacophony of chital, sambar and langur calls we arrived at one corner to find Noon walking nonchalantly with a chital fawn swinging in her mouth.

Noon walks along the shore of the lake. A few peahens take flight. She is heading for a bank of high grass. Suddenly two peacocks fly out of the grass closely followed by two racing cubs. The male, in the lead, grabs the fawn and darts back to a clump of grass. Noon licks her female cub, and they both recline at the edge of the grass. The male polishes off most of the carcass but towards the end Noon interrupts him. He snarls at her, but she ignores him and picks up the remnants.

The male cub moves away. The female cub joins her mother and they chew on the bits and pieces the male cub has left. An hour has gone by and soon mother and cubs leave the grass, walking across the edges of the lake towards Padam Talao. The cubs jump and chase each other as they walk around the lake. One charges into Noon and she snarls in annoyance. All three have now reached the edge of the first lake. The cubs turn a corner and disappear around the edge of the lake.

Noon walks in front of us on the vehicle track. Two sambar alarms blast the silence near the cubs. They have been seen. Noon is now completely alert and darts forward on the track, realizing that the sambar is caught between the track and her cubs. The cubs are assisting her unintentionally. There is a thud of hooves and a noise in the undergrowth. Noon has settled down on her belly, frozen to the ground at a point where a narrow animal path leads out from the edge of the lake. She has judged the exit point exactly. A large rock hides her. In a flash she leaps into the forest and is out of sight. We hear a grunt.

Moving ahead, we find her a few metres off the track, in the throes of killing an adult sambar hind. She has a perfect killing grip on the throat. The sambhar's legs twitch in vain: the grip is firm. The cubs approach cautiously and watch her intently. The male cub moves to the carcass but a flicker from the sambhar's hind leg forces it to retreat. In minutes the sambar is dead. The male cub rests his forepaw on the rump, while Noon still holds her grip and the female cub stands near her mother. Noon drags the carcass away to where a thick bush makes visibility difficult. The cubs jump all over the sambar and their mother. A kill of this size feeds mother and cubs for three days.

Later that evening we found two of Nalghati's cubs posing for us on a bunch of rocks. As the sun set and I was getting ready to go I found my jeep wouldn't start and I soon realized that there was no petrol in the engine. The tigers watched us from a few feet away and the moment had to come to get out of the jeep and walk back. Luckily as I was about to step out another jeep arrived and I got into it, leaving behind my own jeep with two tigers sniffing at it!

Later that month, I had returned from a morning drive and had a brief glimpse of Noon and her cubs as they moved towards a dense area below the walls of the fort. I was eating breakfast when a sambar alarm called twice. I left everything, grabbed the closest camera and, with Fateh's son, jumped into my jeep. As we came to the clearing between the two lakes we found Noon and her cubs moving across towards Rajbagh. She had suddenly changed her mind about her day shelter.

It is 10.30 in the morning. The setting is perfect as the three tigers cut across the first lake, with the backdrop of the fort. Soon they pass an old ruin that must have been the entrance to a mosque in times gone by. Now overgrown by grass and shrubs, it is one of Noon's regular day shelters. She passes by, but her male cub, curious and frisky, decides to investigate and climbs up the steps to a parapet where he poses against the minarets. A splendid sight. He watches us for a while before jumping down to follow Noon. A sambar in the distance bellows in alarm, a peacock takes flight, shrieking as the family moves towards Mori, a secluded corner of the second lake. They quench their thirst around a small pool of water. One of the cubs jumps right over it and they slowly walk away and out of sight. The few sambar calls at breakfast had been responsible for a fruitful encounter and I feel cheerful as I turn the jeep to head back to base.

Suddenly one of the largest sambar stags I have ever seen came galloping out of the area into which the tigers had disappeared. It was closely pursued by Noon.

I could not believe it. Did Noon expect to kill such a large stag in a chase over open ground? Stag and tigress were out of sight some 30 metres ahead. Fumbling with the starter of the jeep I moved ahead, heart pounding. A couple of metres from the vehicle track in the clearing the sambar stag stood motionless. Noon was clinging on to the side of its neck. Her canines had a grip, but they were nowhere near the throat. Both tiger and sambar were frozen in this position, staring at each other. There was not a sound or a movement. I watched aghast.

Goverdhan, Fateh's son, woke me from my daze: 'Come on, use your camera, we will never see this again.' My reverie broken, I swung into action, cursing myself in the process. In the rush of departure I had picked up the nearest camera which had only an 85 mm lens, and no more than twenty shots left; I had no more film on me.

Such is jungle life sometimes. I took a few quick pictures, not sure how steady my hands were in the excitement of the moment. Goverdhan warned me to be careful. Sambar and tiger were still locked together. I decided to change position and move up to within 3 metres of them. They were not in the least bothered by our presence, too involved in their own struggle. I was so close that I felt I could touch them. I chose my shots carefully so as not to finish the film too soon. Noon seemed unwilling to shift her grip, and allow the sambar to escape. A few parakeets flew overhead, some green pigeons chattered in a tree nearby as tiger and sambar remained locked together.

In a few minutes the sambar, with a great heave of his neck, shrugs the tigress off, but in a flash she attacks his forelegs in an effort to break them. The stag jerks away, but Noon goes again for the neck, rising on her haunches with one paw on his shoulder for leverage. But in vain. The sambar swivels around and Noon now attacks the belly. With much struggling in the region of the belly and hind legs, she succeeds and the sambar finds itself in a sitting position while the tigress takes a firm grip on one of its hind legs. My film is slowly running out. Suddenly Noon's male cub appears and stands motionless, observing the encounter. Noon and the sambar are again frozen in their position. The cub inches closer, perhaps sensing victory. Noon yanks at the hind leg, opening the skin and trying desperately to break it. This is the only way she will prevent the stag from escaping, as her grip is not a fatal one.

Suddenly, the sambar, utilizing every ounce of his strength, shakes Noon off, stands up and runs. The cub flees in fear, and an exhausted Noon tries to chase the sambar. The sambar, with a burst of adrenalin, escapes in the direction from which he had come, towards the edge of Rajbagh. Noon lopes after him, but hasn't the energy to sustain any speed. Her cubs come around as if egging her on but she snarls at them in irritation.

The cubs run ahead, following the sambar to the water's edge. The stag alarm calls for the first time, a strange dull and hollow sound as if his vocal chords had been damaged in the attack. Seeing the approaching tigers he wades into the lake. Noon and her cubs now watch anxiously from the shore. The stag has great difficulty moving through the water. He stumbles forward and finds himself in a patch of deep water; he is forced to swim and nearly drowns. His head bobs up and down, his limbs move frantically as he struggles to reach the far bank. The tigers follow along the shore but Noon soon gives up and reclines at the edge instead, exhausted and panting. Her tongue is cut and bleeding. Her cubs jump around her but she snarls at them and they lay down to rest under the shade of a bush.

The stag limps towards the shore and stands motionless for many minutes in the shallow water. Noon watches for a bit but then decides against pursuing her quarry, walking away into the dense cover of Mori to shelter. Her cubs follow. The sambar slowly hobbles out of the water on to a bank of grass. His right foreleg looks twisted and broken; patches of skin have been raked and a bloody injury swells on the side of his neck. The magnificent stag is now a sorry sight. How he had found the strength to escape I can't fathom.

So much of the tiger's life centres on food that it is around the kill that you can best see and learn about the tiger's habits and behaviour patterns. In November 1987 we had a unique encounter with a transient male and Noon and her cubs.

The male tiger has reached the far side of the grass, 30 metres from Noon's family, and at the edge of the grass he flops down and appears to go to sleep. By this time Noon is carefully inching her way out of the long grass towards the male, her eyes peering anxiously in an effort to scan the area. She suspects possible danger but the grass is too tall for her to pinpoint the position of the intruder. She remains frozen for some minutes, looking carefully in the general direction of the male tiger. Suddenly the male flicks his tail, a few blades of grass move and Noon has located her quarry. Completely alert and tense, she starts to stalk towards him. It takes her fifteen minutes to move 30 metres. She moves so gently that you cannot hear the sound of her weight on dry grass. Head lowered, muscles bunched up near her shoulders, she comes to a halt a couple of metres from the male tiger. Now she remains frozen, watching him, clearly realizing his potential threat. She looks as if she will attack. The male has his eyes closed. I wonder if he is playing a game.

Cautiously Noon takes another step forward and suddenly the male swivels around, confronting her with a vicious growl, and in a flash both tigers rear up on their hind legs, literally standing to face one another. They keep their balance for a bit as they gently try and slap each other, then lower themselves with the most blood-curdling growls. They rise on their hind legs three times in this way before sitting down to face each other. The forest goes silent. But finally the male has his way and moves into the grass to feed on Noon's kill.

Noon is up and quickly tries to return to her carcass. In the distance I see her male cub running away. The male tiger follows Noon. The female cub now leaves the grass and attempts to nuzzle the male but he snarls and mother and daughter watch him as he enters the grass and appropriates the carcass. Noon follows him but as she approaches, she is met by a series of low growls. She retreats with her cub to settle down at the edge of the grass in the shade of a tree. There is no sign of the male cub. After a while the female cub attempts to enter the grass, approaching cautiously. She persists until she is within a metre of the male but his aggression soon forces her away.

In the last years of the 1980s, Fateh got transferred out of Ranthambhore and much against his wishes was posted as Director of Sariska Tiger Reserve— but he could never be as deeply involved in Sariska. Those were difficult years for me. I lost both my parents in 1987 and much of the next two years were spent just absorbing the loss. I spent much time in Ranthambhore, a lot of energy going into the establishment of an organization that could create more harmony between the people and the Park. And our first effort was to provide primary health care. In 1989, the Ranthambhore Foundation started its community conservation activities.

The Nineties : Crisis Years

It must have been February 1990. I had just bought some land near one of the villages that faced the Park. My project was to look at how arid landscapes could be re-greened. Only the year before I had initiated the Ranthambhore Foundation by starting a Mobile Health Service.

One evening on the agricultural fields near my land I heard a strange shrieking noise. Rushing across with a torch, I found a tiger fighting with a wild boar. As the tiger reared up on its hind legs, the boar tried to charge at its belly. They roared and grunted at each other and I remained frozen watching this extraordinary spectacle. The tiger was no match for the big tusker and within minutes it retreated.

For the first time, young tigers were spilling outside the Park and were seen around the villages. The National Park was fully occupied and young tigers were being forced outward. I remember once even seeing a tigress with three cubs walking the edge of the Park late in 1990, in a region sparsely covered with trees. But it was also in 1990 while I was busy diversifying the activities of Ranthambhore Foundation, that there was endless talk of tiger poaching.

From 1992 started our years of horror. One of the first horrors we faced was the 1992 tiger census in which the Ranthambhore Foundation participated. Day and night, for a week, we searched for signs of tigers. We couldn't believe the population of tigers had crashed within a couple of years. Indications were that Ranthambhore was left with about fifteen tigers, from the peak of fifty in the 1980s. For a year there had been persistent reports of missing tigers and the sightings had gone down. Fateh over the last few years had become exceedingly worried about a few tigers that he thought were missing. He would rush back from the Sariska sanctuary and find that he couldn't locate some tigers he knew. I remember him dashing off a letter to the Chief Wildlife Warden complaining that tigers were going missing. I was engrossed in my work outside the Park. I was not ready to even consider that poaching could take place. Despite rumours of poaching however, no one had been able to believe that the superb tigers of Ranthambhore could be killed by man. Our complacency was unforgivable.

The findings of the 1992 census convinced me that all the rumours of poaching that I had ignored were true. Frustratingly, the government not only rejected our census figures but also the entire census operation; it claimed that a ten-minute drizzle had ruined the soil strait which is crucial when enumerating tigers. (The paws of the tiger leave prints on the soil, and by counting the different pug-marks it is possible to know the total number of tigers in a certain region.) But these were all excuses and the reality was about to strike.

A few months later a poacher was caught with a tiger and a leopard skin just on the edge of Ranthambhore. After the police interrogation of the poacher we realized that at least fifteen to twenty tigers had been killed by poachers over the last two years. The tiger tragedy at Ranthambhore only reflected what was happening across India. As tiger populations had declined in South-east Asia, the poaching mafia across the world had zeroed in on India to fulfil the demands of the new and horrific market that had opened up in tiger bones and skins. They are used as magical cures for arthritis and rheumatism, and are converted into wines for sexual potency. Ever the tiger's penis has a price and is cooked as soup to provide sexual powers. The success of efforts in reviving the population of the Indian tiger meant that it had a price on its head. We already had our hands full with India's enormous human population and severe habitat loss. Now tiger poaching had hit us as well.

We had to change our strategies to fight this development. So far we had never participated directly in the actual governance of wildlife. I was anonymous, carefree and without a connection to government. But the disaster of Ranthambhore forced some of us to get involved with the government.

Along with others, I have now served on every government committee dealing with tigers and wildlife. We have tried to network with non-governmental organizations, encouraging new strategies. Meanwhile the seizures of tiger skins and bones have been depressingly regular. There has been much government indecision, apathy, indifference and the inability to convert recommendations into field actions. The big NGOs have appalling squabbles—the politics of conservation within and outside government reflect the tragic state of affairs in the conservation world.

In 1993 I advised again on the census as a part of the central government since I had been inducted into the Steering Committee of Project Tiger. I reached a conclusion of fifteen or twenty tigers in Ranthambhore. Few tigers could be seen. They had retreated again under their nocturnal cloak. The Ranthambhore National Park was being destroyed not just by poachers; camps of graziers were to be seen everywhere and the very precious and sparse grass of Ranthambhore was grazed to exhaustion by cows and buffaloes. Prey populations were declining.

In 1994 the Supreme Court petition ordered that an expert group look into the numbers of tigers in Ranthambhore. My colleagues and I sat in a hall full of hundreds of plaster-of-Paris casts of pug-marks and a forest staff sold tigers to us like fish in a fish market. Even colleagues who swore by the pug-marks, and claimed responsibility for the methodology were confused. Each of us had different assessments. The methodology had fallen apart. Some of it was even being faked.

The delight of simply watching tigers and recording their natural history has been replaced by hours spent dealing with an ever-increasing crisis. It has also been a politically fragile period and governments have changed so quickly that serious field strategies have suffered tremendously. The issue of saving tigers is not a political priority. In the last seven years thousands of square kilometres of the tiger's habitat have either been degraded or lost forever, and an increasing number of large-scale projects, like mines, dams, power projects and industry have exacted their price. It is as if there are a million fires and mutinies in the land of the tiger, and we are running around trying to douse the fires, but for each one doused ten new ones start.

In the 1990s, I worked as closely as I could with five different Ministers of Environment and Forests and one Prime Minister. In the process, I learnt the answer to a question often asked: 'Why can't India save tigers?' The answer is long and complicated. But in essence it lies in the fact that India never really created a mechanism for proper governance of forests and wildlife. Because of this the tiger is in grave danger. I am at present writing a book on this whole issue and what the experience of the 1990s was for me and so many others.

he Future

We are in the middle of 1999 and the tiger is still alive. India probably has 2000 to 2500 tigers, nearly 50 per cent of the world's tiger population. We also have one billion people on the eve of the millennium. In 1992, after the tragedy that hit Ranthambhore's tigers, I was certain that by 1999 the tiger would be virtually extinct. Luckily, I have been proved wrong. Though there has probably been a steady decline of 200 tigers each year, against all the odds, there is still a wild population alive. The animal has a remarkable ability to survive against extreme pressure.

The levels of problems and pressures will now be extraordinary. Even today the central and state structures are cracking up under the pressure of the problems as more and more people, business, projects and so much more compete for forest land and the treasures they hold. India is in the worst ever recession that it has known and some people believe that because of actual bankruptcy and the bankruptcy of ideas, our economy is on the verge of collapse. As elsewhere in the world we must be alert to the big international and multilateral organizations, which advocate the exploitation of natural resources to overcome ever increasing international debt.

Sadly, the economic crisis has forced the government to freeze recruitment. As a result, the number of trench soldiers of the forest, the forest guards, has been reduced by 60 per cent since none of the vacancies have been filled. The forest guards look after the most vital treasures of a country and these natural resources will determine the future of the country and the life of its people. New vehicles and other essential equipment cannot be easily purchased because of the fund-crunch.

We lose at least 10,000 square kilometres of forest each year and at least 50,000 square kilometres of dense forest canopy are thinned out by the timber mafia. The central government puts the loss each year from forests at 50,000 crores or 1200 million dollars. The actual level of depletion could be far higher. Few realize that 300 rivers and perennial streams flow from the forests of the tiger. Natural resources are exploited by politicians and businessmen; the exploitation is then rationalized in the name of 'development'. It is a partnership in crime that cripples this country each day as the face of our land is ripped apart and scarred; rivers are drowned in so many toxic chemicals that river water has become a lethal cocktail of poison. Environment and wildlife laws are violated and every structure of enforcement is paralysed. Resources allocated to protect our environment add up to 50 crores each year when our loss is 50,000 crores annually.

Let us go back to Ranthambhore and look at some of the problems it faced, which reflects the state of wildlife governance in India. Between 1992 to 1996 three field directors were posted there in quick succession; they had no wildlife training or experience. They were being sent to run one of the finest tiger habitats in the world without any knowledge of tigers, wildlife or ecology. This is a problem plaguing the entire country and all our protected areas. And it afflicts all ranks of officers that serve a protected area. Sometimes you can have six senior officers all without any knowledge about managing wildlife.

I remember visiting the Park with my colleagues on four different occasions between 1994 and 1997. On one occasion, I was with the senior-most forest officer of India, the Inspector General of Forests, and early one morning we were standing in the heart of the forest, surrounded by cows, buffaloes and endless masses of cow dung. Graziers a hundred yards away were yelling abuse at us. The Inspector General was shocked, the forest officers indifferent. I wondered bitterly if our conservation venture in Ranthambhore needed to be renamed 'Project Buffalo'.

What we require immediately is a sub-cadre within the Indian Forest Service that is trained and experienced in wildlife management. The Indian Forest Service administers 20 per cent of India's land mass; half of it is arid land that requires replanting and re-greening. Part of the other forested land is still used for commercial timber operations. And the remaining 4 per cent is in our protected area system that requires a new Indian Wildlife Service to manage it. The Ministry of Environment is too overworked to do anything about such matters. It spends much of its time clearing large-scale projects in relation to environmental issues and the problems of pollution. The time has come to separate functions. There must be a Ministry of Environment and a Ministry of Natural Resources (Forests and Wildlife).

Disbursement of funds too is a problem at Ranthambhore. Nearly thirty-seven government offices and officers have to sign a single file before the funds reach the field. The central government needs to create a Protected Area Authority of India that has a bank account and is able to receive money from federal and private sources and then disburse this money directly and quickly to the field.

It could also deal with other problems. For example, a national survey carried out in the middle of the 1980s revealed that 69 per cent of the protected areas surveyed had human populations and 64 per cent had community rights, leases or concessions inside them. Grazing occurred in 69 per cent of the surveyed protected areas; in 57 per cent of the areas, there was collection of non-timber forest produce. Ranthambhore in the last decade has seen the population grow at 3 per cent each year, increasing to an enormous 125,000 in a 3-kilometre radius from the Park boundary. Domestic livestock has increased by 6 per cent each year and today 60,000 cows, buffaloes, sheep, goats and camels try to enter the Park to graze. All protected areas of India have the same problem. In order to minimize the impact of these pressures it is vital to involve local communities in the protection of the area and work with them on different ideas. We must try and create green belts outside the protected area that deflect, absorb and minimize the pressures on the forest. The Ranthambhore Foundation has attempted to do just that for the last ten years. Forest departments must try and do the same.

Since its inception, no serious research has been undertaken on predators, prey or habitat in Ranthambhore. Management decisions are not based on science and arbitrary decisions have always plagued the Park. Science, field research and training must form a vital part of national policy on wildlife. Creation of a new Ministry, a new sub-cadre in the Indian Forest Service and a Protected Area Authority are vital first steps for the future.

However, even in the current tragic scenario there are rays of hope. One of best things to happen was the remarkable recovery in Ranthambhore in 1997 when a forest officer, G.V. Reddy, got to do the job that he joined the forest service for. A new and dynamic Director of Project Tiger, P.K. Sen, and the Field Director of the Tiger Reserve, Rajeev Tyagi, helped the process and a new Chief Wildlife Warden for Rajasthan, R.G. Soni, strengthened the resolve to revive Ranthambhore.

G.V. Reddy confronted for the first time, illegal graziers and wood cutters, enduring their aggression. He was also the first senior officer to discuss with local villagers and NGOs strategies to protect the Park. Under him, the tragic apathy of government over many years seemed to have faded. By 1998, four tigresses and their cubs were flourishing and I spent a delightful month again watching tigers in Ranthambhore. On one single day, I even saw nine tigers. What a difference the right man in the right job can make!

This does not mean that the crisis is over, only that action needs to be focused and urgent. P. K. Sen had commented:

It is my considered opinion, after more than one and a half year as Director, Project Tiger, that the tiger and its ecosystem is facing its worst ever crisis. I feel the figure of one tiger death everyday may even be an underestimate and there are many reasons to say so. If out of every ten tigers poached, poisoned or crushed under the wheels of a vehicle, three are tigresses who have cubs, all the cubs will also die unnoticed because they are totally dependent on the mother. The death of three resident male tigers will result in new males occupying vacant ranges and in the first instance they will kill all cubs in order to father their own litters. Thus for every ten tigers killed, sometimes as many as fifteen additional tigers die, since this entire process completely disturbs the natural life cycle of the tiger. Such is the devastation we face.

The gravity of the situation is evident when we look at the recent Forest Survey of India report where the tiger state of Madhya Pradesh has lost nearly 4000 square kilometres of forests and 13,000 square kilometres of dense canopy. The situation in Bihar, Andhra Pradesh, Assam, the North East and parts of Madhya Pradesh and Karnataka is as depressing. It is not only that we are desperate for funds for the management of tigers and their ecosystem, but urgent restructuring and reforms are also required. We must re-prioritize these issues in both central and state governments by creating new ways for time-bound and effective field action, both in the disbursement of money and translating new policy into the field. Unless these measures are implemented, I fear that soon after the turn of the century our tigers, the very symbol of our natural heritage, will be relegated to a handful of protected areas in India.

On the occasion of 25 years of Project Tiger, unless revolutionary steps are taken immediately, there is little hope for the future and we could be reaching a point of no return.

In 1999 Fateh and I spent another month observing, from morning to night every day, the superb tigers of Ranthambhore. The Park has continued to recover under the leadership of G.V. Reddy. An old friend who was District Collector of Sawai Madhopur in the 1980s, Salauddin Ahmed, had become Principal Forest Secretary, and the team putting Ranthambhore on the right track was further strengthened. Fateh was appointed Honorary Wildlife Warden. We couldn't believe Ranthambhore's luck. How long would it last?

I remember early one morning in February watching an area between the first two lakes as the winter mist slowly rose. Suddenly a tiger started roaring from a hundred metres behind me. In seconds, from around a kilometre away, came back two different responses from other tigers. Later that day, Fateh and I watched two sub-adult tigers feed with their mother up an incline of a hill.

In 1999 there were at least six instances of tigress and sloth bear interactions and in all these cases the sloth bear chased away the tigress after they had both reared up their hind legs as if to box each other. These are unique encounters in the natural world. There have been reports of sloth bears being killed by large male tigers in Panna Tiger Reserve and more recently a large male tiger was seen eating the carcass of a sloth bear in Kanha. But to witness the interactions in Ranthambhore was a delight.

1999 has been very special because for the first time, I saw a Ranthambhore I had never seen before. At noon I was driving around Rajbagh to investigate a sambar carcass when I saw a ratel eating off it, watched a few metres away by a jackal that it had chased off. Shy, elusive and nocturnal, it was off in a flash the moment it spotted my jeep. I had seen a new species after a quarter of a century of watching. It just shows what the forest hides and how each day here is a day of discovery—an ongoing process of learning more about the magic of the wilderness...

It is June 1999. We are waiting impatiently for the skies to darken and bring in the last monsoon of the twentieth century. We are also anxiously awaiting signs of a litter from at least three of the tigresses that were resident in and around the lakes. Another century is about to start in the lives of the wild tigers of Ranthambhore.

The rains crash down in early July and by the end of the month we hear that forest guards have spotted two tigresses that have tiny litters.

Valmik Thapar

The eyes of a tiger caught in a flash of light - it is at such moments that the power and sheer beauty of this animal is completely mesmerizing.

This is tiger time. The last rays of the setting sun catch a tiger as it starts its evening stroll.

A feeding tiger caught in the headlights of a jeep. In the seventies, when this was taken, such glimpses were rare.

A tiger walks by four peacocks that carefully watch its passage. They are an excellent indicator of the presence of tigers, shrieking out in alarm when sensing one.

As a tiger walks past some peahens one of them takes off in alarm. Interpreting the sounds of birds is vital when locating tigers.

 ▲ A tiger in the grass, which is excellent camouflage while hunting .

A tiger sleeps, while at least 8 frogs jump around on its side and shoulders.

A tiger cub rolls on its back just after a meal.

▲ An 18 month-old cub folds its massive paws as it sprawls across the ground.

A tiger sits alert.

A 16 month-old cub snarls down from a rock. This is when their first permanent teeth start to appear.

Alarmed sambar deer watch the passage of the tiger (Noon) at the edge of the lake.

A tiger in the foreground with a small group of sambar in the back. The sambar are not concerned as the tiger is some distance away.

Two 16 month-old cubs wait at dusk for the return of their mother who is out searching for prey.

A tiger patrols its area

Tiger portraits

One of the biggest male tigers we have ever recorded in the forests of Ranthambhore.

Noon dozes, watched by a small blue kingfisher that is waiting to dive into the pool of water. I watched the two for many hours on this hot summer's day.

Noon and her cub ➢
watch from the cover
of high grass.

▲ Kublai, a male tiger, watches alertly.

▼ Man-made as well as animals paths have to be patrolled and home ranges marked. Tiger can walk more than 15 km. in a night

An early morning encounter with mother and cub.

A tiger pads along the road at dusk. Our jeep follows in reverse.

A 16-month-old cub watches two peacocks from the tall grass. ➢

A 24-month-old sub-adult races across shallow water chasing a white-necked stork.

A 20-month-old cub attempts to stalk a vulture at the edge of the lake.

A tigress snarling at the photographer.

The last rays of the sun vanish and a tiger stalks the forest.

Three tigers and a peacock. Cubs learn the art of hunting by practising on peacocks, hares, and other small animals.

A perfectly concealed tiger watches a spotted deer, waiting for the right moment—but it is too far away.

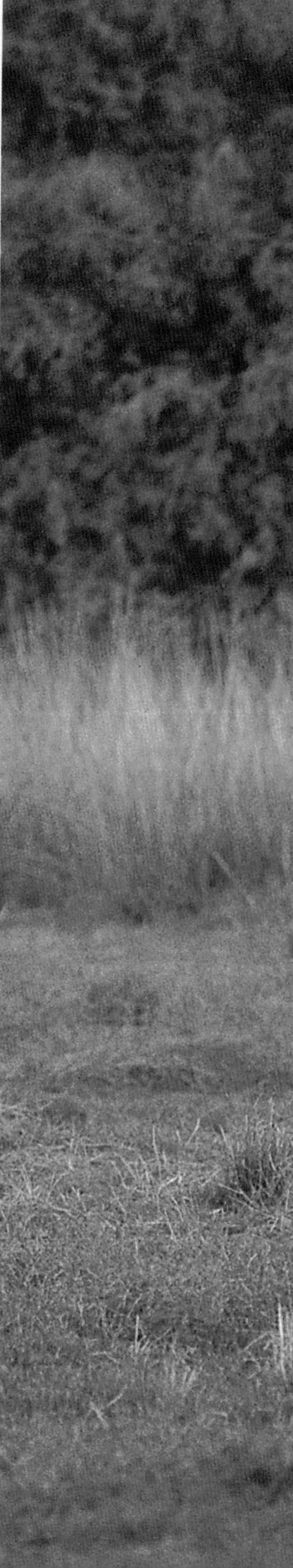

Noon stalks along the edge of grass step by step, as if in slow motion, in search of her quarry.

Success at last—just after a spotted deer kill.

One of the 18 month-old cubs rushes off with the carcass of a spotted deer fawn and mother and siblings race around to cut off escape. But usually the dominant cub gets away with small prey and the rest will wait till it is finished with the meal. Like *buzkashi* with tigers.

⋀ Two tigers around the carcass of a nilgai—a crow sits warily on it, tempted by the possibility of scraps.

⋁ Padmini drags the half-eaten carcass of a nilgai. Nine tigers fed on this carcass.

Noon carries a small chital back to her two cubs: when prey is small it dangles from her mouth as she carries it, the legs of the victim leaving drag marks on the forest floor which indicate the event. Sometimes drag marks can go on for even 2km.

Noon brings her cubs to the kill. The crows indicate the position of the carcass. When she leaves the kill she ensures that it is well-hidden from scavengers by camouflaging it with grass, leaves and mud, which she scrapes over it with her paws.

A tigress choking a langur monkey to death. Langurs only fall victim to tigers when they wander off from trees to the ground. Observing tiger attacks on langurs is very rare.

A tree-pie indicates the exact point in the grass where a tigress lies with her kill. These birds are excellent indicators of a tiger's presence, especially when it has a kill.

 Four tigers just after feeding on a spotted deer in the dry deciduous forests of Ranthambhore.

Genghis watches a sambar stag and hind from the edge of the lake. The deer are unaware of the tiger's presence.

Genghis races across the edge of the lake to create confusion among the feeding deer. He hopes that in the chaos he will be able to attack the most helpless deer.

A tiger races out from the grass to cut off the escape of a sambar.

Genghis chases into the water after sambar and targets a young fawn that is unable to move quick enough against the weight of the water. One out of ten tries are successful in the water. This is a much better average than on dry land where it is one out of fifteen or twenty.

▽ The chase continues as a pair of black-headed ibis fly off. The tiger uses its powerful limbs more effectively in shallow water than the sambar can.

Noon attempts to bring down an enormous sambar stag. With her hind legs providing leverage her powerful forelegs knock a 600 lb. sambar stag to the ground. The weight of the tigress is probably 200–230 lb.

Once the stag is down, Noon clings to its hind legs. One of its cubs appears, convinced that the sambar is down, and the feast can commence.

Crocodiles attack a sambar hind in the waters of the lake. Under the water they pull and tear at the limbs, snout and ears. It is a slow death for the deer. Crocodiles can take hours to kill a sambar. A tiger comes by to watch and then settles down on the bank. It waits patiently for a chance to scavenge the carcass, watching every move.

Racing into the water and pounding the crocodiles away with a huge splash, a tigress heaves the carcass out amidst a crowd of angry but helpless crocodiles. Tiger–crocodile interactions are unique to Ranthambhore. Most of the time the tiger gets the better of the crocodiles.

As the tigress pulls the carcass out, one of her cubs tries to help by pulling at its mother's tail.

▲ A young tigress guards her kill of a sambar stag. Kills are zealously protected against scavengers.

▼ Feeding in thick cover is what tigers prefer to do; it prevents them from being bothered by crows, vultures and other scavengers.

An adult male tiger, Broken Tooth guards the carcass of a sambar that he has appropriated from the resident female. The tigress has been forced off her kill.

A tigress relaxes against the low lying trunk of a tree as she protects her chital stag carcass. She will protect it for days, feeding off it even after the meat has putrefied and been infested by flies and maggots.

A tiger straddles the carcass of a sambar stag, dragging it to a cool thicket. Tigers can drag enormous weights, sometimes more than double their own.

Noon in the process of killing a sambar female. Her cubs watch alertly. One of them pats the sambar with its paw. It is a process of learning and the tigress is an expert teacher.

A rare moment—Noon and Kublai mating under a flame of the forest tree. A sambar carcass floats in the water in front of them. Between bouts of mating the tigers eye the carcass that they will feed on later. After mating the female literally slaps the male off her as she swivels around.

A tigress changes dens when disturbed, carrying her cubs gently in her mouth from one point to another. She is ruthlessly protective and will not tolerate any intrusion in the area around her den.

A young cub, a few weeks old, nuzzles its mother. A tigress remains completely devoted to her cubs for nearly two years.

▲ Laxmi suckles her cubs—this is the age when milk is supplemented by a diet of meat and the tigress will either bring back meat to her cubs or take them to it.

▼ Two cubs entangled with each other as if wrestling—again this is how they will develop their forelegs into formidable weapons.

A large cub caught in the air just as it is about to land on its sibling who is quenching its thirst at a water-hole.

A tigress boxes out at a male to prevent him from appropriating her kill, and threatening her cubs. He wins and forces the tigress and her cubs to retreat.

Aggression in tigers can take place over mating, territory and food. It can also happen when a new male wants to father his own litter and tries to get rid of the small cubs of the previous male. Here a male tigers slaps at a female.

An 8 month-old cub atop a tree as its mother Noon passes below.

A cub chases past its mother, towards its sibling. Keeping the cubs in control requires instilling discipline into them. This is vital if the mother is to hunt with any success.

A cub races up the trunk of a tree as its mother watches. As long as the cubs are light they move up and down trees with the agility of leopards. After 16 months they will be too heavy for tree-climbing.

Noon watches as her two cubs swat each other. These bouts of playful behaviour are crucial to learning how to use their powerful limbs.

▲ At 18–20 months, cubs have endless bouts of wrestling and boxing and here their mother reveals her anger and irritation. This is the moment the process of breaking apart starts
▼ within the family unit. They are too big now to tolerate such close physical proximity.

At 18 months the cubs have started marking trees and bushes with their scent. Here two cubs make an effort to mark a tree while the third sniffs at the bark.

Claw marking is a frequent activity of tigers—here a male reaches high up on a trunk to leave its territorial signal. These signals and spray marking control conflict between tigers. Tigresses in oestrus will attract male tigers through their scent marking.

A tigress sits before a chatri, built 200 years ago in memory of a courageous warrior.

Using the chatri as an observation post, or even as a day shelter is a frequent activity of Ranthambhore's tigers.

A tigress fast asleep inside a chatri at Rajbagh palace.

A tiger wades through the shallow waters of Rajbagh.

Ranthambhore is at the extreme western end of the tiger's territory. In summer the leafless trees provide excellent visibility that allows for long and close observation of tigers.

 The dry deciduous forests of Ranthambhore, typical habitat for tigers.

A tigress rests as a herd of cheetal graze peacefully in the background. Sometimes predator and prey appear to be relaxed in each other's presence. ➢

A tigress in front of Padam Talao, the lotus lake. Ranthambhore fort and Jogi Mahal make a spectacular backdrop in which memories of man merge with nature.

 Two tigers explore the receding waters of Padam Talao. It is the beginning of the monsoon.

A tigress looks up at the tallest cliffs in Ranthambhore. Scores of vultures nest here.➢

⩓ A tiger cools off in the waters of Rajbagh lake. A black-headed tern flies overhead.

⩔ Amidst painted storks, sambar deer watch a tiger in the distance.

▲ A tigress approaches a gaggle of grey lay geese, winter visitors to Ranthambhore, that fly in from Siberia. The bird life on these lakes is rich and diverse.

A young tiger looks back, soon after the arrival of the last monsoon of the twentieth century.

The tiger is not the only predator that walks these forests. The leopard stalks the upper reaches keeping as far away as it can from the tiger. Here a leopard has been 'treed' by another leopard while warring over the favours of a female.

A sloth bear digs up the earth in search of food. A peahen passes by and a tree-pie waits to pick up the odd insect.

Sloth bears share this land of the tiger and frequently chase away tigers that they encounter.

If a fawn stumbles into the shallow water a crocodile will grab it quickly. Here the mother watches her fawn in the crocodile's jaws. The crocodile will feed only when the flesh has putrefied and it can tear it apart.

The Indian marsh crocodile is unable to attack large sambar deer especially on dry land.

Crocodile and fawn.

A sloth bear walks by two peacocks.

▼ The king vulture is an excellent indicator of the tiger. A pair of them get ready to take off. They are most vulnerable to a tiger's attack when on the ground.

The jackal is able to carry off the kill even though this deer is nearly double its size.

A jackal attempts to defend its spotted deer kill from a crow.

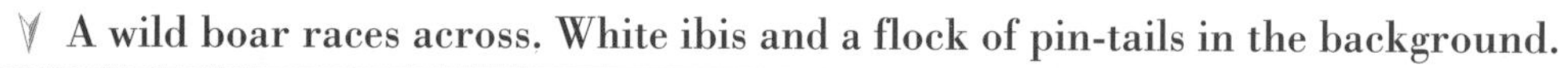

A wild boar races across. White ibis and a flock of pin-tails in the background.

The only time in 25 years that I could take pictures of the rarest of Ranthambhore's predators—the caracal—one of the most endangered cats of the subcontinent.

A leopardess sneaks back to eat a spotted deer she had killed. A hyena had annexed it from her, earlier.

A leopard has sneaked into the tiger's domain and killed a sambar. It feeds warily knowing that it will be forced off if a tiger appears.

Fateh Singh Rathore on a tragic day in the 1980s with three 9month-old cubs killed by a territorial male tiger, in an adjacent forest tract.

Tigers also die when they are poisoned over a livestock kill. A dead tiger has been found by a forest patrol, floating down a water-body just outside the National Park.

APPENDICES

Counting Tigers in Ranthambhore
K. Ullas Karanth

Ranthambore's tigers have gained fame among wildlife aficionados of the world through several books and films produced since the 1980s. Valmik Thapar, Fateh Singh Rathore, the late Kailash Sankhala and others have, pictorially and in prose, captured the beauty of these tigers in their majestic setting. These popular accounts of Ranthambore's tigers have helped to focus the world's attention on problems relating to their conservation, and to that of conserving tigers elsewhere in the world. However, to translate this global goodwill into effective action, we also need to monitor tigers and their prey using reliable science.

As a conservation scientist working with the Wildlife Conservation Society (WCS), my research, therefore, has focused on understanding the tiger's ecology in varied habitats across India. Starting from my long-term study site in Nagarahole in Karnataka, I have studied tigers and prey species in several protected areas in India: Pench, Kanha, Kaziranga, Namdapha, Sundarbans, Bhadra and Bandipur. As a part of these studies, I have tried to develop scientifically reliable methods to count tigers. One such method, which is particularly useful in areas where the big cats occur at relatively high population densities, is known as 'camera trapping' or 'photo-trapping'.

The patterns in which stripes are arranged on the head, body and limbs of any tiger, are unique to that individual. By matching these stripe patterns with those on other individuals using photographs, it is possible to reliably identify each individual tiger. If stripes on the tiger's flanks are to be compared, it is necessary to compare the same side (either left or right) among the different animals, because the stripe patterns differ between the two flanks of the same animal.

However, to be able to compare stripes, I need clear photographs of tigers taken at relatively close quarters. The 'camera traps' are devices used for obtaining such close-up photos of tigers. A camera trap comprises a small transmitter located on the side of any trail used by tigers. It emits an electronic beam, aimed at a receiver placed on the opposite side of the trail, about 7 metres away. The receiver is connected to two cameras capable of automatically setting the exposure, focusing the picture and rewinding the film. When a tiger, or some other large animal, walks along the trail and interrupts the electronic beam, the two cameras fire and take its pictures from both the sides.

With adequate effort at camera trapping, I am able to gradually build up a photo-library of several individual tigers using an area in which the cameras are placed. However, the science of wildlife biology has long recognized a critical problem in counting animals. Whatever the method used to count animals, in most situations it may not be possible to count all individuals in a population. In other words, a 'census' which means a 'total count' is rarely feasible. Often, only some individuals out of all animals in a population are counted. Over the years, wildlife biologists, mathematicians and statisticians have worked on this problem and come up with several ways of 'sampling' animal populations to estimate their numbers.

In studies in which animals are marked and counted, such as bird banding, rodent trapping, and tiger photo-trapping, an estimation method called 'capture-recapture sampling' is used. Essentially this method uses the frequencies in which identified tigers turn up in repeated 'samples' of the population, to estimate the proportion of animals from the total that are actually photographed. Therefore, I am able to estimate the numbers of tigers without photo-capturing all the tigers. Such population estimates become more reliable; if more tigers use the area, the sampling effort is greater, and an estimation model which fits the data better is selected.

With the encouragement of the Director of Project Tiger, Rajastan Forest Department and the non-governmental organization Tiger Watch, I decided to include Ranthambhore Tiger Reserve among my tiger survey sites. Camera trapping was done in Ranthambhore during May and June 1999. My research team deployed cameras at 60 sampling points in Ranthambhore (p. 145) to achieve a sampling effort of about 840 camera-trap-nights. Besides tigers, several other interesting and even rare animals turned up in the photographs. During the sampling effort at Ranthambhore, 16 different individual tigers were photo-captured (pp. 146-7). The estimated average number of tigers, derived using the best fit capture-recapture model was 20 tigers within the effectively sampled area of 244 square kilometres. Thus, the estimated average tiger density in the sampled area works out to 8.2 tigers per 100 square kilometres. My earlier studies in high quality tiger habitats such as Nagarahole, Kanha and Kaziranga show densities ranging from11–17 tigers per 100 square kilometres. Since the Ranthambhore tiger population is probably still recovering from adverse biotic impacts in the recent past, it may not yet have attained its full potential density. However, detailed studies of densities of prey species are needed to support this conjecture.

Caught in a camera-trap, the ratel is the most elusive of Ranthambhore's nocturnal animals.

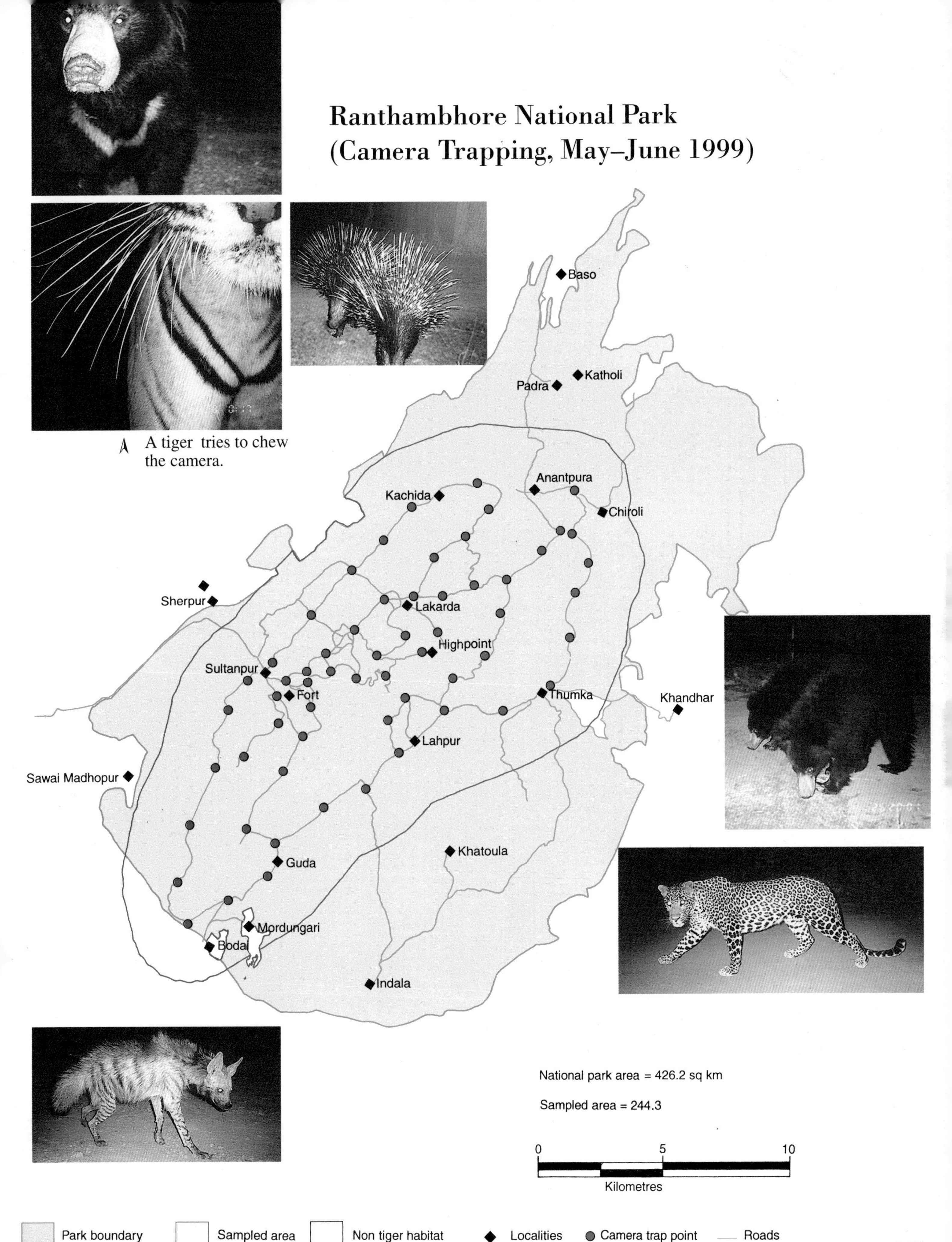

A tiger tries to chew the camera.

16 different tigers captured by 'Camera Traps' in Ranthambhore

Interpreting Ranthambhore's Forests

Using Remote Sensing by satellite Ranthambhore Foundation has sponsored the Regional Remote Sensing Service, Jodhpur, to interpret the forest cover over an area of 620 sq. km. which includes the area of Ranthambhore National Park, every two years from 1992.

Category	Area in hectares			
	1992	1994	1996	1998
Good tree cover or biomass	13,471.40	12,526.87	13,903.88	14,823.48
Moderate tree cover or biomass	10,797.23	11,536.00	11,614.36	9.810.20
Less tree cover or biomass	14,128.34	15,623.28	14,119.14	13,506.38
Forest blank	23,332.40	22,205.27	22,265.67	22,852.48
Others	330.11	163.06	156.43	372.47

Looking at these statistics it is very clear that from 1992 (a year when serious levels of poaching of tigers and timber were detected) the forest lost some of its good tree cover, and in 1996 some improvements started that continue to sustain themselves into 1998. Even in the less tree cover category there are improvements from 1994 to 1998. Some of these improvements in the habitat have also revealed a better state for the tiger since it was in early 1998 that the recovery of Ranthambhore's tiger population started.

Remote sensing of the forest cover is vital especially when you can check in the field and 'ground-truth' some of the data.

The most positive observation of 1998 reveal improvements in the National Park Area and with that there has been a spread of water bodies and some new ones have formed. The water regime improvement is an excellent indication of the health of the habitat and vital to the future of wildlife.

Please note: 1997 figures of tigers have not been confirmed by the Directorate of Project Tiger because of the methodology followed. Counts of tigers are based on the pug-mark methodology that suffers from a margin of error.
Stop Press: It is understood that Satpura National Park , Bori Sanctuary and parts of Pachmarhi Sanctuary will be India's 26th tiger reserve and Nameri in Assam and Pakhui in Arunachal will be the 27th 'inter-state' tiger reserve.

S.No	Year of Creation	Name of Tiger Reserve	State	Total Area	Tigers
1	1973-74	Bandipur	Karnataka	866	75
2	1973-74	Corbett	Uttar Pradesh	1316	138
3	1973-74	Kanha	Madhya Pradesh	1945	114
4	1973-74	Manas	Assam	2840	125
5	1973-74	Melghat	Maharashtra	1677	73
6	1973-74	Palamau	Bihar	1026	44
7	1973-74	Ranthambhore	Rajasthan	1334	32
8	1973-74	Similipal	Orissa	2750	98
9	1973-74	Sunderbans	West Bengal	2585	263
10	1978-79	Periyar	Kerala	777	N.R.
11	1978-79	Sariska	Rajasthan	866	24
12	1982-83	Buxa	West Bengal	759	32
13	1982-83	Indravati	Madhya Pradesh	2799	15
14	1982-83	Nagarjunsagar	Andhra Pradesh	3568	39
15	1982-83	Namdapha	Arunachal Pradesh	1985	57
16	1987-88	Dudhwa	Uttar Pradesh	811	104
17	1987-88	Kalakad-Mundanthurai	Tamil Nadu	800	28
18	1989-90	Valmiki	Bihar	840	53
19	1992-93	Pench	Madhya Pradesh	758	29
20	1993-94	Tadoba-Andheri	Maharashtra	620	42
21	1993-94	Bandhavgarh	Madhya Pradesh	1162	46
22	1994-95	Panna	Madhya Pradesh	542	22
23	1994-95	Dampha	Mizoram	500	5
24	1998-99	Bhadra	Karnataka	492	_
25	1998-99	Pench	Maharashtra	257	_
			TOTAL	33875	1458

S.No.	Name of State	1972	1979	1984	1989	1993	1997
1	Tamil Nadu	33	65	97	95	97	62
2	Maharashtra	160	174	301	417	276	257
3	Kerala	60	134	89	45	57	NR
4	West Bengal	73	296	352	353	335	361
5	Orissa	142	173	202	243	226	194
6	Karnataka	102	156	202	257	305	350
7	Bihar	85	110	138	157	137	103
8	Assam	147	300	376	376	325	458
9	Rajasthan	74	79	96	99	64	58
10	Madhya Pradesh	457	529	786	985	912	927
11	Uttar Pradesh	262	487	698	735	465	475
12	Andhra Pradesh	35	148	164	235	197	171
13	Meghalaya	32	35	125	34	53	NR
14	Manipur	1	10	6	31	_	NR
15	Tripura	7	6	5	_	_	NR
16	Mizoram	_	65	33	18	28	12
17	Nagaland	80	102	104	104	83	NR
18	Arunachal Pradesh	69	139	219	135	180	NR
19	Sikkim	_	_	2	4	2	NR
20	Gujarat	8	7	9	9	5	1
21	Goa, Daman & Diu	_	_	_	2	3	6
22	Haryana	_	_	1	_	_	NR
	TOTAL	1827	3015	4005	4334	3750	3435

Fact Sheet for Ranthambhore Tiger Reserve

1.1 It comprises Ranthambhore National Park, Sawai Man Singh Sanctuary, Keladevi Santuary and adjoining reserve forest area.
1.2 Area 1334 sq. kms.
1.3 Zoning

Core Zone (National Park)	274.50 sq.km.
Buffer Zone	118.0 sq.km.
Sawai Man Singh Sanctuary	127.60 sq.km.
Keladevi Santuary	674.00 sq.km.
Reserve Forest Area	140.54 sq.km.

1.4 Dates of establishment

Sawai Madhopur Wildlife Sanctuary	1955
Project Tiger	1973
Ranthambhore National Park	1980
Sawai Man Singh Sanctuary	1984
Keladevi Sanctuary	1984

1.5 Location
Reserve area spreads between: Latitudes 25 46' N to 21 12' N,
Longitudes 76 17' E to 77 13' E
Nearest town : Sawai Madhopur
H.Q. : Sawai Madhopur
1.6 Topography

Average elevation	: 350 mts. M.S.L.
Highest Point	: 540 mts. M.S.L.
Lowest Point	: 215 mts. M.S.L.

1.7 Climate
Temperature varies from a summer-highest of 48°C to a Winter-lowest of 2°C.
Annual Rainfall: 800mm. Rainy Season: July to September
1.8 Forest type: Dry deciduous and dry thorn forest. Most of the areas come under *Anogeissus pendula* forest.
1.9 Indicative fauna: Tiger, Leopard, Sloth Bear, Hyena, Caracal, Ratel, Sambar, Chinkara, Crocodiles, etc.
The Reserve is rich in avifauna. It has a checklist of more than 256 species including a variety of raptors, water birds and gallinaceous species.
1.10 Tourism period : October to June.
1.11 Administrative structure: The Field Director is the overall incharge of the area. The Reserve area is divided into two parts/divisions, core area and buffer area. These divisions are managed by two Deputy Field Directors.

Ranthambhore Foundation Initiatives

Around Ranthambhore National Park and across India

Objectives:

The long-term objective of the Ranthambhore Foundation is, first, the maintenance of the essential ecological balance necessary for man to live in harmony with nature—in the Sawai Madhopur district of Rajasthan, and in rural and forest communities in other parts of India; and second, to undertake every possible measure necessary to ensure wildlife and forest conservation, especially *protection of the tiger and its habitats* all over India and in other tiger range countries.

Description of Activities:

1. Health care, medicine, and family planning programme: A mobile health care and family planning programme has been in operation since 1988. Under this programmeme, a mobile medical unit comprising a doctor and other medical staff regularly visits some 15 villages every week. A team of health workers working in different villages also facilitates family planning and other services with the help of the mobile unit. Approximately 50,000 people benefit from the programme. A Primary Health Centre has also been in operation since 1997. The health service is now implemented by the Prakratik Society.

2. Tree nursery, re-greening, seed collection, and afforestation: A major programme for planting trees in the area has been in operation since 1989. A mother nursery of 65,000 saplings is being maintained to supply the villages. Every year after the rainy season, tree planting campaigns are organized. A seed bank is also maintained to collect and preserve seeds for sapling germination. Several green satellites in different villages have already been created. A complete record of tree planting and growth rate by species is being documented and maintained.
3. Promotion of dairy Development and animal husbandry: The dairy development and animal husbandry programmed aims to introduce and propagate crossbred animals in the area, promote milk cooperatives for effective marketing and provide veterinary services. Artificial insemination and natural service facilities are also provided to the villagers. A dairy demonstration and breeding centre was also established ten years ago; it is to be phased out by January 2000 in the hope that local organizations will sustain such endeavours.
4. Promotion and propagation of appropriate technology and alternative energy: Encouraging the use of bio-gas, thereby minimizing the use of fuelwood.
5. Informal nature education programme: A programme to educate local youth and children about the environment, forest, and wildlife, and their underlying interrelationships.
6. Community organization and local groups: From the beginning, a concerted attempt has been made to facilitate community organization and local bodies or groups that are educated about the need to conserve nature and to empower them to take up programmes for environmental protection. Several such groups are now actively working for forest and wildlife conservation in the area. The Ranthambhore Foundation supports local organizations with a view to strengthening them. By the year 2000 we will have restructured our field activities in order to play a more catalytic role, not just around Ranthambhore but across India.
7. Support and assistance to Central and State Governments has been an essential part of our activities. We have tried to help them in getting jeeps, motor cycles, trucks etc. for patrolling and anti-poaching activities, and have supported a series of awards for courage and bravery of forest staff across India. The Executive Director of the Foundation serves on all the National Committees that advise governments on wildlife policy and tiger conservation. We have also printed and distributed *In Danger*, a book in English and Hindi on India's wildlife, in order to spread the awareness of India's endangered wildlife. To this end, we have also been organizing for the past four years, a monthly lecture or film show at the India International Centre, New Delhi.
8. Tiger Link Network: The Delhi office of the Foundation recently proposed the concept of a network for tiger conservation at an all-India level. This network is being coordinated by the Foundation and number of organizations working in various parts of the country for tiger conservation have as a result come together for effective and coordinated field actions in their respective areas. Tigerlink attempts to strengthen existing field actions and promote new initiatives wherever necessary. The objective is to mobilize all resources and strengths and launch a collective effort to save the tiger in India's natural forests. The Tiger Link Network is growing fast, with more and more individuals and organizations coming into the fold. It is now well-recognized that if tigers cease to exist, several other species, and even the forest system itself, will face the same fate in the near future.

At present, there are over one thousand participants in Tiger Link internationally. Annual meetings of the Link at the national, and three regional levels, are organized to facilitate evaluation and assessment of the field actions and work out effective plans for action. Tiger Link newsletters are published thrice a year by the Foundation on behalf of the network and are sent to all the Links and partners. The main office of the Link in Delhi acts as a nerve centre and coordination office for the network, maintaining an information centre and database for the benefit of the network participants on all vital issues and statistics. It is now proposed to establish three regional coordination centres for better and effective coordination and networking among the links.

The Ranthambhore Foundation hopes to restructure and reshape its activities to make them more effective, innovative, economical and strong. There are enormous pressures on the forests and tiger habitats of this country and any effort to minimize negative impacts will play a critical role in the future of the tiger, all wildlife and indeed, the planet itself. We hope that in the twenty-first century we will continue to play a catalytic role in this process.

Contacts: Valmik Thapar/Sunny Philip/Dr G S Rathore
Tel: +91-11-301-6261 Fax: +91-11-301-9457 E-mail: tiger@vsnl.com
Address: Ranthambhore Foundation, 11 Kautilya Marg, Chanakyapuri, New Delhi 110021

Special Facts

1. Ranthambhore Fort, nearly one thousand years old and several kilometres in circumference, is open from dawn to dusk and is a very special place for bird-watching, occasionally leopards use the area. Small water-holes and ponds are found all over the area. Dramatic views of the lakes and the forest below give a feeling of the scale of the park. Occasionally tigers can also be seen from the top of the fort.

2. The forests of Ranthambhore are dotted with old ruins, tombs, tanks and step-wells. The fort of Khandar on the far side of the Reserve is well worth a visit. These imposing forts command stunnig views of the area. To visit some of these historic sites—Khemsa Kund, Rajbagh, Jogi Mahal, Kualji temple and others—in and around the Reserve, contact the office of the Deputy Field Director of the core area for special permission.

3. Gilai Sagar, Mansarovar and several other waterbodies in and outside the reserve attract thousands of migratory birds during the winter. All these places are exciting to see especially for bird-watchers. Contact the Deputy Field Director's office for more information and permission.

4. Two river systems, the Chambal and the Banas are within an hour of the Tiger Reserve and can provide an interesting glimpse of riverine ecology.

5. The Ranthambhore Foundation and World Wide Fund for Nature (WWF) India run community conservation schemes on both sides of the National Park. Tiger Watch, another NGO works closely with the Park administration to strengthen field management. For those interested in observing some activities contact local co-ordinators of these organizations for permission. Dastkar, another NGO, runs a craft centre that has revived traditional skills and Prakratik Society runs a health centre and an eye clinic to deal with primary health care. The Ranthambhore School of Art Society is renowned for the skills of it's wildlife painters.

6. In Ranthambhore there has been only one case of a tiger killing a boy and dragging him into a 'nallah' where a chunk of flesh from the thigh was consumed. It happened on a day when 200,000 people walk through the National Park to receive the blessings of Lord Ganesh. At dusk a boy went off towards some shrubs to relieve himself and was pounced on by a tiger who was probably desperate to reach a water-hole amidst all the traffic.

More recently in May 1999 a leopard picked up a two-year-old girl inside Sawai Madhopur town near a cement factory and dragged her to an abandoned boiler-room, where he ate her leaving only the head behind. The room was finally stormed by forest guards and the snarling leopard tranquillized, though he died while being transported to the Kota zoo.

Normally tigers and leopards do not enter Sawai Madhopur town (1 km. away) and only once was a tiger seen walking the streets of Sawai Madhopur and in the garden of Sawai Madhopur Lodge—the Taj group hotel. Dazed and lost it was finally tranquillized at the edge of the road and released back inside the forest.

Tigers, bears and leopards have mauled at least 16 people in these 25 years and the level of man–animal conflict is surprisingly low considering the close proximity of town and villages to the border of the National Park.

Addresses and telephone numbers (telephone area code: 07462)
- Field Director - Tel: 20223
- Deputy Field Director - Tel: 21142
- Ranthanmbhore Foundation, near village Sherpur. Tel: 520039, 20286
- Prakratik Society, near village Sherpur
- WWF India, Khandar, Sawai Madhopur
- Tiger Watch - Tel: 20811
- Dastkar, near village Kutalpura
- Ranthambhore School of Art Society, Ranthambhore Rd, opp. Hotel Ranthambhore Regency

FURTHER READING

Ahmad, Yusuf S. *With the Wild Animals of Bengal* (Y.S. Ahmad, Dhaka, 1981)
Alfred, J.R.B. et al. *The Red Data Book of Indian Animals, Part 1: Vertebrata* (ZSI, Calcutta, 1994)
Ali, Salim. 'The Mohgal emperors of India as naturalists and sportsmen' (*Journal of the Bombay Natural History Society* 31 (4): 833–61, Bombay)
Alvi, M.A. and A. Rhaman. *Jahangir–The Naturalist* (Delhi, 1968)

Baikov, N.A. *The Manchurian Tiger* (Hutchinson, London, 1925)
Baker, S. *Wild Beasts and their Ways* (London, 1890)
Barnes, Simon. *Tiger* (Boxtree, London, 1994)
Baze, W. *Tiger, Tiger* (London, 1957)
Bedi, Rajesh and Ramesh Bedi. *Indian Wildlife* (Brijbasi, New Delhi, 1984)
– *Wild India* (Brijbasi, New Delhi, 1990)
Biscoe, W. 'A tiger killing a panther' (*Journal of the Bombay Natural History Society* 9(4): 490, Bombay,1895)
Brandar, A. Dunbar. *Wild Animals of Central India* (Arnold, London, 1923)
Breeden, Stanley and Belinda Wright. *Through The Tiger's Eyes: Chronicle of India's Vanishing Wildlife* (Ten Speed Press, USA, 1997)

Campbell, T. 'A tiger eating a bear'*(Journal of the Bombay Natural History Society* 9(1): 101, Bombay,1894)
Chakrabarti, Kalyan. *Man-eating Tigers* (Darbari Prokashan, Calcutta, 1992)
Champion, F. *In Sunlight and Shadow* (Chatto & Windus, London, 1925)
– *With a Camera in Tiger Land* (Chatto & Windus, London, 1927)
Choudhury, S.R., Khairi. *The Beloved Tigress* (Natraj 1999)
Corbet, G.B. and J.E. Hill. *The Mammals of the Indomalayan Region* (Natural History Museum/OUP, London/Oxford, 1992)
Corbett, G. 'A tiger attacking elephants' (*Journal of the Bombay Natural History Society* (7)1: 192, Bombay,1892)
Corbett, J. *Man Eaters of Kumaon* (Oxford University Press, Oxford, 1944)
Courtney, N. *The Tiger–Symbol of Freedom* (Quartet Books, London, 1980)
Cronin, E.W. *The Arun: A Natural History of the World's Deepest Valley* (Houghton Mifflin, Boston, 1979)
Cubitt, Gerald and Guy Mountfort.*Wild India* (Collins, London, 1985)

Daniel, J.C. *A Century of Natural History* (BNHS/OUP, Bombay, 1986)
– *A Week with Elephants* (BNHS, Bombay, 1996)
– *The Leopard in India–A National History* (Natraj, Dehra Dun, 1996)
Davidar, E.R.C. *Cheetal Walk* (OUP Delhi, 1997)
Denzau, Gertrude and Helmut. *Konigstiger* (Tecklenborg Verlag, Steinfurt, 1996)
Desai, J.H. and A.K. Malhotra. *The White Tiger* (Publications Division, Ministry of Information & Broadcasting, New Delhi, 1992)
Dharmakumarsinhji, R.S. *Reminiscences of Indian Wildlife* (OUP, Delhi, 1998)
Divyabhanusinh. *The End of a Trail–The Cheetah in India*, (Banyan Books, New Delhi, 1996)

Eisenberg, John F., George McKay and John Seidensticker. *Asian Elephants–Studies in Sri Lanka*, (Smithsonian, Washington, 1990)

Fend, Werner. *Die Tiger Von Abutschmar* (Verlag Fritz Molden, Vienna, 1972)
Flemming, Robert L.Jr. *The Ecology, Flora and Fauna of Midland Nepal* (Tribhuvan University, Kathmandu, 1977)

Gee, E.P. *The Wildlife of India* (Collins, London, 1964)
Ghorpade, M.Y. *Sunlight and Shadows* (Gollancz, London, 1983)

Green, M.J.B. *IUCN Directory of South Asian Protected Areas* (IUCN, Cambridge, 1990)
– *Nature Reserves of the Himalaya and the Mountains of Central Asia* (OUP, New Delhi, 1993)
Gurung, K.K. *Heart of the Jungle–the Wildlife of Chitwan, Nepal* (Andre Deutsch, London, 1983);
– *Mammals of the Indian Sub-continent and Where to Watch Them* (Indian Experience, Oxford, 1996)

Hanley, P. *Tiger Trails in Assam* (Robert Hale, London, 1961)
Hardy, Sarah B. *The Langurs of Abu* (Harvard University Press, Cambridge, USA, 1977)
Hillárd, Darla. *Vanishing Tracks–Four Years Among the Snow Leopards of Nepal* (Elm Tree Books, London, 1989)
Hooker, J.D. *Himalayan Journals*, 2 vols (John Murray, London, 1855)
Hornocker, M., *Track of the Tiger* (Sierra Club Books, 1997)

Israel, S. and Toby Sinclair. *Indian Wildlife* (Apa Publications, Singapore, 1987)
Ives, Richard. *Of Tigers and Men* (Doubleday, New York, 1995)

Jackson, Peter. *Endangered Species–Tigers* (The Apple Press, London, 1990)

Khan, M.A.R. *Mammals of Bangladesh* (Nazima Reza, Dhaka, 1985)
– *The Handbook of India's Wildlife* (TTK, Madras, 1983)
Krishnan, M. *India's Wildlife 1959–70* (BNHS, Bombay, 1975)

Littledale, H. 'Bears being eaten by tigers' (*Journal of the Bombay Natural History Society* 4(4): 316, Bombay, 1889)
Locke, A. *The Tigers of Trengganu* (London, 1954)

Manfredi, Paola. *In Danger* (Ranthambhore Foundation, New Delhi, 1997)
Mattiessen, Peter. *The Snow Leopard* (Chatto & Windus, London, 1979)
Mcdougal, C. *Face of the Tiger* (Andre Deutsch and Rivington Books, London, 1977)
McNeely, A. Jeffrey and P.S. Wachtel. *The Soul of the Tiger* (Doubleday, New York 1988)
Menon, Vivek. *On the Brink: Travels in the Wilds of India* (Penguin India, New Delhi, 1999)
Meacham, Cory. *How the Tiger Lost Its Stripes* (Harcourt Brace, New York, 1997)
Mishra, Hemanta and Dorothy Mierow. *Wild Animals of Nepal* (Kathmandu, 1976)
– and Jefferies, M. *Royal Chitwan National Park; Wildlife Heritage of Nepal* (The Mountaineers, Seattle, 1991)
Montgomery, Sy. *Spell of the Tiger* (Houghton Mifflin, Boston, 1995)
Morris, R. 'A tigress with five cubs' (*Journal of the Bombay Natural History Society* 31(3): 810–11, Bombay, 1927)
Moulton, Carroll and Ernie J. Hulsey. *Kanha Tiger Reserve: Portrait of an Indian National Park* (Vakils, Feffer & Simon Ltd., Mumbai, 1999)
Mountfort, G. *Tigers* (David and Charles, Newton Abbot, 1973)
– *Back from the Brink* (Hutchinson, London, 1978)
– *Saving the Tiger* (Michael Joseph, London, 1981)
Mukherjee, Ajit. *Extinct and Vanishing Birds and Mammals of India*, (Indian Museum, Calcutta, 1966)
Musselwhite, A. *Behind the Lens in Tiger Land* (London, 1933)

Niyogi, Tushar K. *Tiger Cult of the Sundarbans*, (Anthropological Survey of India, Calcutta, 1996)
Naidu, M. Kamal, Trail of the Tiger (Natraj, Dehra Dun, 1998)

Oliver, William. *The Pigmy Hog* (Jersey Wildlife Preservation Trust, Jersey, 1980)
Owen Edmunds, Tom. *Bhutan* (Elm Tree Books, London, 1989)

Panwar, H.S. *Kanha National Park–A Handbook* (CEE, Ahmedabad, 1991)
Perry, R. *The World of the Tiger* (Cassell & Co. Ltd., London, 1964)
Philips, W.W. *A Manual of the Mammals of Sri Lanka*, Parts 1–3 (Colombo, 1980–84)
Prater, S. *The Book of Indian Animals* (BNHS, Bombay, 1988)

Ranjitsinh, M.K. *The Indian Blackbuck* (Natraj, Dehra Dun, 1990)
– *Beyond the Tiger, Portraits of South Asian Wildlife* (Brijbasi, New Delhi, 1997)

Richardson, W. 'Tiger cubs' (*Journal of the Bombay Natural History Society* 5(2): 191, Bombay, 1890)
Roonwul, M.L. and S.M. Mohnot, *The Primates of South Asia*, (Harvard University Press, Cambridge, USA, 1977)

Saharia, V.B. *Wildlife in India*, (Natraj, Dehra Dun, 1982)
Sanderson, G.P. *Thirteen Years Among the Wild Beasts of India* (W.H. Allen & Co., London, 1896)
Sankhala, K. *Tiger* (Collins, London, 1978)
Schaller, G.B. *Mountain Monarchs–Wild Sheep and Goats of the Himalaya* (Chicago University Press, Chicago, 1977)
– *The Deer and the Tiger* (Chicago University Press, Chicago, 1967)
– *Stones of Silence; Journeys in the Himalaya* (Andre Deutsch, London, 1980)
Scott, Jonathan. *The Leopard's Tale*, (Elm Tree Books, London, 1985)
Seidensticker, John, Sarah Christie and Peter Jackson. *Riding the Tiger: Tiger Conservation in Human-dominated Landscapes* (Cambridge University Press, 1999)
Seidensticker, John, *Tiger* (Voyager Press, 1996)
Shah, Anup and Manoj. *A Tiger's Tale* (Fountain Press, Kingston-upon-Thames, 1996)
Shahi, S.P. *Backs to the Wall; Saga of Wildlife in Bihar* (Affiliated East-West Press, Delhi, 1977)
Sharma, B.D. *High Altitude Wildlife in India* (Oxford & India Book House, New Delhi, 1994)
Sheshadri, B. *The Twilight of India's Wildlife* (John Baker, London, 1969)
Singh, Billy Arjan *Tiger Haven* (Macmillan, London, 1973)
– *Tiger Haven* (OUP, Delhi, 1988)
– *Tara, A Tigress* (Quartet Books, London, 1981)
– *Tiger! Tiger!* (Jonathan Cape, London 1984)
– *The Legend of the Man Eater* (Ravi Dayal, New Delhi, 1993)
– *Tiger Book* (Lotus Roli, Delhi, 1997)
Singh, K. *The Tiger of Rajasthan* (London, 1959)
– *Hints on Tiger Shooting* (The Hindustan Times Ltd., Delhi, 1965)
Skaria, Ajay. *Hybrid Histories: Forests, Frontiers and Wildness in Western India* (OUP, New Delhi, 1999)
Stebbing, E.P. *Jungle By-ways in India* (London, 1911)
Stracey, P.D. *Tigers* (Arthur Barker Ltd., London, 1968)
Sukumar, R. *The Asian Elephant* (Cambridge University Press, Cambridge, 1989)
– *Elephant Days and Nights* (OUP, New Delhi, 1994)
Sunquist, Fiona and Mel. *Tiger Moon* (University of Chicago Press, Chicago, 1988)

Thapar, Valmik. *With Tigers in the Wild* (Vikas Publishing, Delhi, 1983)
– *Tiger:Portrait of a Predator* (Collins, London, 1986)
– *Tigers: The Secret Life* (Hamish Hamilton, London, 1989)
– *The Secret Life of Tigers* (OUP, New Delhi, 1998)
– *The Tiger's Destiny* (Kyle Cathie, London, 1992)
– *The Land of the Tiger* (BBC Books, London, 1997)
– *Tiger: Habitats, Life Cycle, Food Chains, Threats* (Wayland Publishers, 1999)
Tikader, B.M. *Threatened Animals of India* (ZSI, Calcutta, 1993)
Tilson, R.L. and V. Seal, *Tigers of the World: The biology, Bio Politics, Management and Conservation of an Endangered Species* (Noyes Publications, New Jersey, 1987)
Toogood, C. 'Number of cubs in a tigress' litter', (*Journal of the Bombay Natural History Society* 39(1): 158, 1936)
Toovey, J. (ed.) *Tigers of the Raj*, (Alan Sutton, Gloucester, 1987)
Tyabji, Hashim. *Bandhavgarh National Park*, (New Delhi, 1994)

Ward, Geoffrey C. *Tiger-Wallahs*, (Harper Collins, New York, 1993)
– *The Year of the Tiger*, (National Geographic Society, Washington, 1998)
Zwaenepoel, Jean-Pierre. *Tigers*, (Chronicle Books, San Francisco, 1992)

For further information, contact:
Ranthambhore Foundation, 19 Kautilya Marg, Chanakyapuri, New Delhi 110 021;
or
Tiger Watch, Maa Forestry Farm, Ranthambhore Road, Sawai Madhopur, Rajasthan.